THE GLOBAL WAR IN UKRAINE

2021-2025

ROBBIN LAIRD

SLDinfo.com

ISBN: 979-8-9989369-4-4

Library of Congress Control Number: 2025918326

Book cover was generated by an AI program.

Second Line of Defense

Arlington, Virginia

To Professor Zbigniew Brzezinski.
Remembering him as a scholar and a teacher.
And thanking him for letting me write my dissertation on a general theory of historical change.
It did not make much sense in a narrow professional sense.
But it made a great deal of sense in terms of my lifetime intellectual odyssey.
I wish I could share this book with him.

CONTENTS

FOREWORD BY
PASQUALE PREZIOSA

The Global War in Ukraine 2021–2025 is a book about foresight, structure, and consequence.

Robbin Laird does not present the Ukraine conflict as a regional aberration; he demonstrates that it is the culmination of unresolved tensions first mapped in the 1990s, when Ukraine and Belarus were identified as "awkward states", buffer spaces too vital to be left alone, too contested to be settled by institutional habit. The lesson that runs through this volume is unsentimental: what the post–Cold War order declined to resolve has returned as a system-defining war.

Laird reconstructs the prelude to the 2022 invasion as an escalatory narrative hiding in plain sight. Russian behavior around Sea Breeze 21, the July 2021 doctrinal framing, and the historical essay questioning Ukraine's legitimacy were not isolated signals but steps in a political-military script, managed escalation, selective de-escalation, tactical opportunism, typical of twenty-first-century authoritarian powers. Set against the optics of the Afghanistan withdrawal and the November 2021 U.S.–Ukraine Charter on Strategic Partnership, Moscow read a combustible mixture of weakness and provocation. The point here is not monocausal explanation but how misread signals, strategic amnesia, and institutional complacency converged to make war more likely.

The central analytic theme of the book is to privilege structure over mood. Putin's approach to NATO, threat inflation anchored in grievance, produced a self-fulfilling security dilemma, treat a distracted alliance as existential, and you will eventually face the unified, militarized NATO you feared.

The result has been transformative. Finland and Sweden entered the Alliance. Germany launched a rearmament that reaches beyond procurement into industrial strategy. The United Kingdom and France have revived a deep bilateral defense compact, including intensified deterrence dialogue and nuclear oversight coordination. The "drone wall" on NATO's northeastern flank acknowledges that air and missile defense is now inseparable from counter-UAS. The Alliance built the NATO Security Assistance and Training for Ukraine framework (NSATU) and federated capability coalitions. And a Europe long on strategic holiday has begun to act as if its security were its own responsibility.

Equally important is Laird's analysis of Russia's wartime political economy. War becomes a mode of domestic control, a self-perpetuating system in which ending the conflict risks unraveling the power structures built around it. Externally, "yuanization" has transformed Moscow's financial operating system: from single-digit shares of trade in 2022 to dominance by 2024, with the yuan becoming primary in settlements and market turnover, an expedient that trades one dependence (on Western finance) for another (on Beijing's policy choices). The vulnerabilities have been visible: Chinese banks curtailing Russia-related yuan transactions under threat of secondary sanctions and liquidity gaps and the realization that reserves and payments now sit at the discretion of another state's central bank and party apparatus. Energy tells a similar story. China buys big, crude and coal volumes are high, but often at significant discounts, turning Russia from price-setter into price-taker. What appears as a pivot is, in practice, a symbiosis with asymmetry: essential for Russia; optional, and therefore leverage-rich, for China.

Laird's global frame avoids clichés. He stresses that China's gains since 2022 are real but qualified. The association with an increasingly isolated Russia complicates relations with Europe, secondary sanctions are a persistent risk, and U.S. controls on semiconductors, AI, and dual-use technologies impose genuine constraints. Moreover, many of Beijing's moves, from Belt and Road to stimulating domestic demand, were underway before the Ukraine war or renewed tariff policies. Opportunism can amplify strategy, it cannot replace it. The broader principle is clear: unilateral disruptions, military or economic, generate second-order openings for competitors that are agile enough to exploit them.

If Russia's coalition is transactional, North Korean shells, missiles, and even troops, Iranian drones and technical cadres, Chinese economic oxygen, Laird shows that the democratic counter-coalition has proved deeper than expected and more innovative than its caricatures. European rearmament is no longer a slogan but a design problem. Germany's defense rethink now spans fifth-generation airpower integration and a wartime industrial partnership with Ukraine itself: Vector reconnaissance fleets, Skynex air defense, TRML-4D sensors, and mass production of strike drones that embody "precision mass" rather than boutique scarcity. Denmark is expanding its F-35 fleet, transitioning air policing, and anchoring Arctic posture while brokering Patriot solutions for Ukraine when U.S. production pipelines were constrained. The Anglo-French renewal is substantive: it reaches into nuclear stewardship and joint oversight precisely because Russia's nuclear blackmail has made European nuclear maturity unavoidable. Parallel to these national arcs, a Coalition of the Willing complements NATO and the EU, with pledged support, joint planning for a future Multinational Force Ukraine, and the political agility to act where consensus bodies move more slowly.

Technology in these pages is not backdrop: it is protagonist. Laird's treatment of military innovation captures the difference between the 1930s and now. Spain was a laboratory for tactics and platforms whose lessons diffused over years. Ukraine is a war lab where cycles of innovation and counter-innovation compress to weeks: drone swarms, AI-assisted targeting, EW-dense environments, software updates as operational tempo. In two years, Ukraine moved from import dependence to near self-sufficiency in drones and wrote the world's first drone-centric doctrine. Operation Spider Web, deep strikes thousands of kilometers into Russia that degraded a significant fraction of its strategic bomber fleet, demonstrated what happens when low-cost airframes, smart guidance, distributed sensing, and audacious targeteering converge. The shift is doctrinal, industrial, and cultural: from exquisite scarcity to intelligent mass, from platform-centric to network-centric, from hardware primacy to software advantage.

The intelligence ecosystem has globalized and commercialized. Alliance Ground Surveillance from Sigonella, U.S. Tritons over the Black Sea, commercial SAR constellations, and the Allied Persistent Surveillance from Space initiative illustrate how state and market have fused to generate persistent ISR. Laird's account of Japan's "eyes in the sky" is emblematic: iQPS and other SAR providers, legal reforms on defense exports, and an emerging space operations posture have moved Tokyo across psychological frontiers, because "Ukraine today may be East Asia tomorrow."

This is how the war in Europe has reformatted Asian security: Japan breaks taboos; South Korea emerges as a pivotal defense producer and standards partner; North Korea becomes a combat actor in Europe, acquiring unwelcome expertise in EW and counter-UAS that will echo on the peninsula. The two Koreas, in different ways, have exported their competition to Europe: Seoul through legitimate industrial partnerships and technology transfer, Pyongyang through illicit supply and expeditionary force contributions.

Laird is equally clear-eyed on the authoritarian axis's limits. Israeli strikes against Iran's nuclear program in 2025 exposed the fragility of Moscow–Tehran cooperation: Russia's muted practical assistance in Iran's hour of need sharpened Tehran's sense of transactional limits even as Russia localized much of Shahed production. Tactical windfalls for Moscow, attention diverted, higher oil prices possible, came with strategic losses: alliance credibility, exposed dependencies, and the reality that authoritarian alignment is a marketplace, not a covenant.

The book delivers another major theme: industrial policy as strategy. By 2025, Ukraine produced a majority of its own military equipment while under fire, an industrial resurrection that validates the resilience of democratic societies when supply chains adapt and allies invest. The West has begun to relearn scale, from Rheinmetall–Anduril cooperation on unmanned systems to broader European rearmament programs. Meanwhile, the "arsenal of autocracy" keeps Russia in the field but at rising cost and diminishing autonomy. The comparison with the 1930s endures, but with a decisive twist: today's interdependence expands coalition breadth, complicates hedging, and makes economic warfare, sanctions, export controls, energy rerouting, capital constraints, central to strategy.

Laird also restores a neglected history: Partnership for Peace. The 1990s and early 2000s contained a remarkable experiment in graduated engagement, practical cooperation, and officer-to-officer trust between Russia and NATO. Forgetting PfP serves narrative convenience on all sides: it impoverishes strategy. Cooperation did not fail because it was naïve; it failed because structural disputes about European order and spheres of influence were left unresolved. Remembering what worked, and why it unraveled, matters for any future architecture that aspires to more than managed confrontation.

No serious account of this war can ignore the nuclear shadow. Drawing on Paul Bracken's "second nuclear age," Laird engages a world where nuclear weapons shape behavior through coercion as much as through the threat of detonation. Sanctuaries emerge; escalation management becomes a head game and the semantic sprawl of "deterrence" undermines policy clarity unless it is specified, who deters whom, from what, with which forces, for how long, at what price. War termination, in such a setting, is not a rhetorical flourish but a design problem: end states, risks, and trade-offs must be faced, not deferred by incrementalism that merely continues yesterday's policy into tomorrow's crisis.

The American dimension is treated comparatively, not polemically. Bush's era of expansive ground commitments; Obama's measured escalations and light-footprint paradoxes; Trump's preference for targeted strikes and de-escalatory off-ramps over large occupations; Biden's completion of withdrawal and extensive but indirect support to Ukraine: each approach carries implications for coalition leadership, burden-sharing, and the credibility of guarantees. The question is not whether diplomacy is possible, but diplomacy from what foundation of power and staying power?

What, then, are the through-lines that this book compels the reader to confront?

There are at least six.

- The first is that structure outlives mood. Geography, ethnicity, industrial base, and great-power rivalry remain stubborn facts. The hopeful assumptions of "inevitable integration" and "peace by trade" collapsed wherever structures were incompatible or contested.
- The second is that coalitions must be built to move, not to admire. NATO, the European Union, and Coalitions of the Willing are not redundant if they address different needs, speed, legitimacy, or execution. Together they provide flexibility, not duplication.
- The third is that industrial policy is deterrence. Political promises without production lines, munitions stockpiles, resilient supply chains, and software talent are little more than performance art.
- The fourth is that technology is changing the grammar of war. Intelligent mass, software advantage, persistent sensing, and human–machine teaming are redefining cost-exchange ratios. Doctrine and procurement must now adapt at the speed of code.
- The fifth is that nuclear coercion is already in play. Strategy must be explicit about thresholds, sanctuaries,

and escalation ladders. In this environment, European nuclear maturity will matter more, not less.
- And the sixth is that memory is a strategic resource. Forgetting the Partnership for Peace, the cycles of failed resets, and the lessons of past cooperation narrows our options and condemns us to repeat familiar errors.

Finally, Laird does not confuse clarity with fatalism. He documents a democratic counter-adaptation that is real: a Europe learning to defend; an alliance system expanding its aperture, an industrial base relearning to scale and a Ukraine that has moved from aid recipient to knowledge provider, doctrinal pioneer, and co-producer. He also leaves open, without illusion, the necessity of a path, someday, for a Russia capable of seeking security through cooperation rather than domination. That possibility does not exist under present arrangements, but it must still be imagined beyond them.

The Global War in Ukraine 2021–2025 offers a map and a mirror: a map of how we arrived in this dangerous present, and a mirror that compels us to confront what will be required to shape a more stable future.

Ukraine is not at the periphery of the international order; it is the furnace in which that order is being reforged, one alliance choice, one industrial line, one software update, and one strategic decision at a time. In drawing these lines together, this volume does more than recount events; it frames a lens for the future.

For that, gratitude is owed to Robbin Laird. He has not only captured a piece of history with clarity and rigor but has also offered a vision of how democratic societies might navigate the trials ahead.

His work reminds us that history is not merely about the past, it is a compass for the future, if we choose to read it.

Lt. Gen. (ret.) Pasquale Preziosa

FOREWORD BY
BRIAN J. MORRA

In Washington, D.C., the foreign policy establishment has struggled to understand the global nature of Russia's war in Ukraine. Fundamental misunderstandings have hamstrung efforts to reinforce Ukraine with weaponry in a timely way and have derailed efforts to influence the numerous countries that provide support either directly or indirectly to Moscow's war effort.

To remedy this, Robbin Laird's new book provides insights that ought to be required reading at the State Department, the National Security Council, the Pentagon, the Congress, and in the Intelligence Community. It will be of equal value to America's allies.

The global nature of the war in Ukraine is further evidence that we are living in a new geopolitical age. Thomas Friedman likens today's chaotic international landscape to a dance floor hosting "a free-for-all square dance of shifting partners".

As Dr. Laird describes in these pages, traditional alignment models have been broken. The old 'east-west' and 'north-south' categorizations so familiar and comforting to foreign policy elites in North America and Europe fail to enlighten us today.

One important manifestation of the new geopolitics is that Asian powers are deeply engaged in the War in Ukraine. The last time Asia was this involved in a European land war was the Mongol invasion in the 13th century.

As Laird shows us, North Korea sends massive amounts of military hardware to Russia and North Korean troops fight alongside Russian forces on the line of contact. Chinese nationals are found on the battlefield and senior Chinese officials have been heard to say that a Russian defeat in Ukraine would be a threat to Beijing, as it would permit the United States to turn its full attention on the Indo-Pacific.

China has a vested interest in a Russian victory or, at least, in a prolonged stalemate. Meanwhile, Japanese and South Korean financial and dual-use aid flows to Kyiv.

Not to be left out, the middle powers – Brazil, India, Turkey, Iran, the Gulf States, and South Africa – are seeking advantage, too. India buys discounted Russian oil, Brazil expands its trade with China which is selling military-grade dual-use technology to Moscow, and Iran is a major military supplier to Russia.

Dr. Laird points out that the institutions undergirding the Western world must deal with an increasingly polyamorous and chaotic world. The starting point is for Western elites to grasp the global nature of Russia's War in Ukraine and not attempt to triage it as an isolated case in today's hyper-interdependent international landscape.

Brian J. Morra

20 November 2025

FOREWORD BY
DR. HOLGER MEY

Robbin Laird's The Global War in Ukraine: 2021-2025 arrives at a critical juncture in contemporary history, offering an essential analysis of a conflict that has fundamentally reshaped the international order. This is not merely a regional war between Russia and Ukraine; it is, as Laird compellingly argues, a systemic conflict that will determine the contours of global security for decades to come.

Understanding how we arrived at this moment requires confronting uncomfortable truths about Western assumptions, Russian imperatives, and the dangerous illusions that permitted this catastrophe to unfold.

The political stability that characterized the Cold War rested on two fundamental pillars: the sobering reality of nuclear weapons and the grudging acceptance of zones of influence. NATO understood the limits of its reach. The Alliance did not intervene when Soviet tanks rolled into East Berlin, Hungary, or Czechoslovakia.

These were painful concessions to geopolitical reality, but they prevented direct superpower confrontation. The signing of the Charter of Paris in 1990 seemingly heralded a new era, one in which the West assumed or perhaps merely pretended that zones of influence had become relics of history, that Russian imperial ambitions had been permanently extinguished alongside the Soviet Union itself.

This assumption proved catastrophically naive. NATO expansion, driven as much by the desperate appeals of former Warsaw Pact nations as by Western strategic design, proceeded throughout the 1990s and 2000s. Russia, weakened and internally focused, could not prevent this eastward movement.

Yet for those who understood Russian strategic culture and historical memory, it was evident that Moscow would neither forget nor forgive what it perceived as Western exploitation of its temporary vulnerability. Russia's elephant memory, as it were, would carry this grievance forward for generations. The question was never whether there would be a reckoning, but when.

Ukraine represented a red line of an entirely different character than Poland or the Baltic states. For Russia, Ukraine has always mattered more than Ukraine has ever mattered to the West. This asymmetry of interest should have been the starting point for any serious Western policy discussion, yet it was consistently over-looked. The possibility of Ukrainian neutrality, a form of Finlan-dization for the twenty-first century, was dismissed almost reflexively. The term itself had become synonymous with appease-ment, carrying connotations of moral weakness and historical failure.

Yet Finland's experience during the Cold War deserves more careful examination than Western policymakers gave it. The Finns mastered a delicate balance, understanding that while it served no purpose to needlessly provoke the Russian bear, they could maintain both dignity and deterrence. Every other day, Finnish leaders reminded the Soviets of the Winter War's brutal arithmetic: for every fallen Finn, ten dead Russians.

This was neither submission nor reckless confrontation, but sophisticated statecraft adapted to geographic and strategic reality. A neutral Ukraine, prosperous and democratic but non-aligned, might have offered a path forward. Whether Russia would have accepted such an arrangement in good faith remains unknowable, because the option was never seriously explored.

The real threat that Ukraine posed to Russia was never military in nature, despite Moscow's rhetoric about NATO encirclement. The existential danger was political: a democratic, prosperous Ukrainian state on Russia's border would have represented a devastating refutation of Putin's authoritarian model. It would have demonstrated to Russians themselves that Orthodox Slavic societies could embrace democracy, rule of law, and European integration successfully. This ideological challenge explains the visceral nature of Russian opposition to Ukrainian independence far more than any security concerns about missile deployments or NATO bases.

Those who dismissed Russia as merely a regional power betrayed a fundamental misunderstanding of both geography and history. When a region extends across eleven time zones and encompasses the world's largest nuclear arsenal, it ceases to be merely regional. From Napoleon to Hitler, underestimating Russia has proven consistently catastrophic. The Red Army stood in East Berlin in 1945 not despite but because of the terrible miscalculation of 1941. Russia in 1917 was on its knees, defeated and revolutionary; twenty years later, Soviet forces were preparing for the titanic struggle that would determine Europe's fate.

When Russia annexed Crimea in 2014, I recall a conversation with a prominent member of the German Bundestag's Foreign Policy Committee. He expressed his disillusionment with Russia, to which I responded that one can only be disillusioned if one harbored illusions in the first place. This exchange encapsulates a broader Western failure: the persistent refusal to understand Russia on its own terms, to grasp its geostrategic imperatives, its historical consciousness, its national character.

I predicted more than thirty years ago that if Ukraine moved decisively toward the West, war would follow. Russia would never accept a Western-oriented, democratic, prosperous Ukraine on its border. This was not prophecy but simple pattern recognition for anyone who seriously studied Russian strategic culture.

Beyond failing to understand Russia, Western policymakers demonstrated an even more fundamental conceptual error: they forgot the political role of military power in international relations. Frederick the Great observed that diplomacy without weapons is like music without instruments. Stalin cynically asked how many divisions the Vatican commanded.

These were not merely clever aphorisms but hard-earned wisdom about how power actually functions in international affairs. When forced to choose between appealing to international courts or commanding superior military forces, states that wish to survive choose the latter. International law matters only to the extent that it aligns with the national security interests of powerful actors willing and able to enforce it.

Future historians will look back with bewilderment at European choices in the post-Cold War era. They disarmed in free fall, systematically dismantling military capabilities accumulated over decades. Then, from this position of self-imposed weakness, they challenged the world's largest nuclear power over a territory Russia considered vital to its core interests. The sheer strategic incoherence of this approach defies rational explanation. It was akin to dismantling fire brigades because fires had become infrequent, then expressing shock when conflagrations rage out of control.

Ukraine is not even the first post-Cold War conflict in Europe. The Yugoslav wars of the 1990s already demonstrated that we had not reached the end of history but witnessed the return of history in its rawest form. Watching those conflicts unfold, I asked a question that seemed obvious but was rarely voiced: What happens when Russia, having endured its period of weakness and humiliation, begins collecting territories and asserting control as it did under the Czars?

The Finns have a saying that captures this permanence: Russia is Russia even if it is fried in butter. Regime changes, economic transformations, and diplomatic engagements may alter Russia's tactics and temporarily constrain its ambitions, but they do not fundamentally change its strategic orientation or sense of imperial destiny.

Two fundamental principles of international relations, understood clearly during the Cold War but somehow forgotten afterward, reasserted themselves with a vengeance.

First, power vacuums do not persist; they are filled, usually by the strongest adjacent power willing to act.

Second, weakness itself constitutes a provocation, inviting rather than deterring aggression.

When Finland and Sweden abandoned their neutrality to join NATO, Russia's response was instructive. Moscow protested but ultimately acknowledged that it could do nothing about Swedish and Finnish membership because these countries fell within the Western sphere of influence. The implicit message was clear: Russia accepted NATO's expansion there because it recognized that territory as beyond its reach, but Ukraine represented an entirely different category, one that fell within Russia's sphere of influence and would be defended as such.

Clausewitz observed that the guilty party in war is always the defender, since the attacker would prefer to advance without resistance. There always exists an alternative to war: raising white flags and surrendering. This brutal logic helps explain Russian calculations. From Moscow's perspective, preventing Ukraine's Western integration justified extraordinary costs, even the deliberate destruction of Ukrainian society and infrastructure. Russia would rather see Ukraine in ruins than prosperous under Western orientation.

This was never simply about territorial acquisition; it was about denying an alternative model of development that threatened the legitimacy of Putin's authoritarian system.

Western statements that Russia must not win and Ukraine must not lose reveal dangerous assumptions about Russian resolve and the risks Moscow proves willing to accept. These declarations presume that Russia, and Putin specifically, would accept defeat in what they have defined as an existential conflict.

But what if this presumption is wrong?

What if Russia proves willing to escalate beyond levels the West finds acceptable?

The conflict becomes a competition in risk-taking, as Thomas

Schelling analyzed during the Cold War. Western nuclear powers Britain, France, and the United States face different risk calculations than non-nuclear Germany. Yet even for nuclear-armed states, the question persists: Would we die for Danzig, as the French agonized in 1939? Would we risk nuclear confrontation over Ukraine?

The West's approach contains an inherent contradiction that reveals confused strategic thinking. Many argue that only NATO membership can provide Ukraine with genuine security guarantees. These same voices warn that if Russia prevails in Ukraine, it will feel emboldened to attack the Baltic states or Poland, both NATO members.

This logic cannot hold simultaneously. Either NATO membership provides effective deterrence and security, or it does not. The reality, uncomfortable as it may be, is that NATO membership alone provides no magical protection. What deters aggression is not treaty language but credible military power and demonstrated political will. A weak, divided NATO invites exactly the challenges its founders sought to prevent.

Russia has nevertheless confronted significant challenges that complicate its strategic position. Moscow fundamentally misunderstood Ukrainian identity and the depth of Ukraine's desire to separate from Russian influence. Russia overestimated its military capabilities, revealing corruption and dysfunction that undermined combat effectiveness. Russia underestimated Western cohesion and the scale of support that NATO members would provide to Ukraine.

These miscalculations imposed severe costs on Russia and transformed what Moscow anticipated as a quick victory into a grinding war of attrition. Yet these setbacks have not altered Russia's fundamental objectives, only the methods and timeline for pursuing them.

Laird's contribution in this volume is to place Ukraine within its proper global context. This is not simply a land war in Eastern Europe, isolated from broader geopolitical competition. It represents a systemic conflict between competing visions of international order, one that links European security to Indo-Pacific stability, that connects NATO cohesion to the credibility of American security guarantees worldwide.

The outcome will shape great power relations for decades, establishing precedents for how territorial disputes are resolved, how nuclear powers interact, and whether international institutions retain any meaningful authority.

What makes this conflict particularly dangerous is the absence of mutually acceptable off-ramps. Both sides have defined the stakes in near-existential terms. Russia cannot accept an outcome that would be perceived domestically as strategic defeat without risking regime stability. Ukraine cannot accept a settlement that legitimizes Russian aggression and leaves it vulnerable to future attacks. The West cannot abandon Ukraine without destroying its own credibility and inviting challenges elsewhere.

Yet none of the parties appears willing to accept the costs and risks necessary to impose their preferred outcome decisively.

This volume provides essential context for understanding how we arrived at this impasse and what the stakes truly are. Laird brings decades of experience analyzing military transformation, alliance politics, and strategic competition to bear on questions that will define our era. His work challenges comfortable assumptions and forces readers to confront difficult realities about power, deterrence, and the limits of diplomatic solutions when core interests collide.

The global war in Ukraine will not be resolved quickly or easily. Its consequences will reverberate through international relations for generations, regardless of how the immediate military conflict concludes.

Understanding this war requires moving beyond simplistic narratives of good versus evil to grapple with the complex interplay of geography, history, capability, and will that shapes great power competition. Laird's analysis provides an indispensable guide for this essential task.

Professor Holger Mey

December 2025

Dr. Holger H. Mey, Honorary Professor for Foreign Policy, Universität of Cologne.

Until mid-2024, Dr. Mey was for 20 years Vice President, Advanced Concepts, Airbus Defence and Space. Before he worked as a consultant to the German Ministry of Defense, a lecturer at various universities, a TV commentator and talk show guest, and op-ed writer. From 1990 to 1990, Dr. Mey served on the Policy Planning Staff for the German MoD. From 1986 to 1990, he was a defence and security policy analyst at the German Think Tank "Stiftung Wissenschaft und Politik". From 1980 to 1986, he studied International relations in Bonn and Los Angeles. Over a period of some 25 years, Dr. Mey gave 86 lectures at the NATO Defense College in Rome and became an Honorary Ancien of the NDC. Dr. Mey still teaches regularly at the NATO School in Oberammergau, Germany, and at the Baltic Defence College in Tartu, Estonia. Dr. Mey has been a member of the IISS for over 30 years and remains a member of the German Council on Foreign Relations in Berlin. He is a member of the Advisory Committee of the Clausewitz Association in Germany.

PROLOGUE

In the early 1990s, as the Soviet Union collapsed and new nations emerged from its ruins, a critical question faced Western policymakers: how should the United States and Europe deal with the newly independent states of Ukraine and Belarus?

These nations, born from the wreckage of communist empire, occupied a precarious position between an uncertain Russian future and an expanding Western order.

Susan Clark's and my study for the Department of Defense's Net Assessment office during this period focused on what we termed "the new Slavic states" – specifically Ukraine and Belarus – and their potential impact on European security. Our analysis, now archived in the Defense Technical Information Center, reached conclusions that have proven remarkably prescient in light of today's ongoing conflict in Ukraine.[1]

As an aside, I should underscore that my relationship with Andy Marshall in Net Assessment was crucial for such a study to even be requested. With a forward leaning thinker like Marshall, there were clients for whom a forward leaning thinker could actually work. It was not a question of aligning your thinking with the Administration of the day and for that I and many other defense analysts owe a lot to "Yoda".

The central finding of the study was stark: these states did not fit comfortably into either the Western or Russian spheres of influence. They represented something more dangerous than mere neutral territory: they constituted potential flashpoints for future European conflict. Ukraine and Belarus existed in a geopolitical limbo that made long-term stability unlikely.

During our field research in the early 1990s, traveling through Kiev, Minsk, and Moscow, we witnessed societies in upheaval. The immediate Western priority was nuclear security, ensuring that Soviet nuclear weapons abandoned in Ukrainian territory came under proper control. But beyond this urgent task lay deeper questions about the future orientation of these nascent states.

The challenge facing Western policymakers was profound uncertainty about how to integrate these "awkward states" into the post-Cold War order. While there was much discussion about the broader fate of Eastern Europe, Ukraine and Belarus presented unique complications. They were neither European in orientation nor willing satellites of Russia.

The West's response was predictably institutional: fall back on default structures like NATO and the European Union as the primary vehicles for managing the new European landscape. This approach worked reasonably well for countries like Poland, Hungary, and Romania, which had clearer Western aspirations. But it left the fundamental question of Ukraine and Belarus unresolved.

Our analysis recognized that any European settlement would ultimately require dealing with Russia, which we "noted that throughout modern European history has challenged or troubled a settled European order." Russian leaders viewed Ukraine and Belarus not as truly independent states but as buffer zones essential to their own security perimeter.

Both Secretary of State Kerry and Vice-President Biden reflected this inability to grasp that the current period of history had not shed the world of spheres of influence, and significantly, failed to grasp how Russia would behave.

Then Vice President Joe Biden in 2011, stated that the United States regarded "spheres of influence as 19th-century thinking," in the context of U.S.–Russian relations. This reinforced the idea that treating neighboring states as a traditional sphere of influence was anachronistic in modern diplomacy.[2]

Then Secretary of State Kerry made the same calculation. On CBS's "Face the Nation," Kerry said: "You just don't in the 21st century behave in 19th century fashion by invading another country on completely trumped up pre-text."[3]

This perspective created an inevitable collision course. As NATO and the European Union expanded eastward over the subsequent decades, they did so without incorporating Ukraine and Belarus. These two states remained suspended between a resurgent Russia and the new European order built after 1991, serving as an unstable buffer zone that we had identified as potentially dangerous to future European stability.

The prescience of our 1990s analysis becomes clear when examining Vladimir Putin's objectives today. Putin's disputes with President George W. Bush during Ukraine's Orange Revolution in 2004-2005 revealed his fundamental goal: to ensure that Ukraine remains like Belarus or dominated by leaders willing to serve as loyal Russian allies rather than pursue Western integration.

Putin's July 2021 essay on "the historical unity of Russians and Ukrainians" and Russia's new military doctrine declaring permanent cultural and political conflict with the West generated in 2021 represent the fulfillment of dynamics we identified three decades ago. The current war in Ukraine is not simply about Ukrainian sovereignty. It is about the stability of the entire post-Cold War European order.

The Russian threat to Ukrainian sovereignty validates our early assessment that these buffer states would become focal points of European instability. Putin's vision of a unified Slavic empire encompassing Russia, Belarus, and Ukraine directly challenges the Western assumption that these nations would naturally evolve toward European integration.

Our early research highlighted the concentration of ethnic Russians within Ukraine and the country's historical oscillation between European and Russian orientation. These internal divisions, identified in the 1990s, continue to complicate Ukraine's path toward consolidated statehood and Western alignment.

Today's crisis demonstrates that the post-Cold War European order remains fundamentally incomplete. The expansion of NATO and the EU, while successful in integrating much of Eastern Europe, left critical questions unresolved about the space between Russia and the West. The vision of a Europe "whole and free" which was the catchphrase from Washington in the 1990s proved more aspirational than achievable.

In fact, a major problem facing American strategic thinking is the love affair with catch phrases, rather than hard debate about the world we actually are living in. It is great to life in the social media "blip world" rather than face the stark realities of a world in conflict and not driving towards an ever greater world peace envisaged by the League of Nations. Oh sorry, I meant the United Nations.

The current conflict represents not an aberration but the playing out of structural tensions identified decades ago. Ukraine and Belarus were always going to be contested spaces because they occupy territory that both Russia and the West consider strategically vital.

Understanding today's crisis through the lens of 1990s analysis reveals several key insights:

- First, the current war transcends Ukraine itself. It represents a fundamental challenge to the post-Cold War settlement and the broader question of whether Europe can achieve lasting stability with Russia as a disruptive power.
- Second, the West's institutional approach to managing post-communist transitions, while successful in many

cases, proved inadequate for the unique challenges posed by the buffer zone states.

- Third, Putin's actions, rather than representing opportunistic aggression, follow a strategic logic aimed at protecting what Russian leaders see as their essential security zone.

The validation of our 1990s analysis offers sobering lessons for contemporary policymakers. The assumption that historical forces inevitably favor democratic development and Western integration has proven naive. Geography, ethnicity, and major power competition remain potent forces in shaping political outcomes.

Resolving the current crisis will require acknowledging the structural nature of the problem we identified thirty years ago. This is not simply a matter of deterring Russian aggression or supporting Ukrainian sovereignty, though both remain important. It requires crafting a new European security architecture that addresses the fundamental incompatibility between Russian and Western visions for the region between them.

The tragedy is that these dynamics were visible and analyzable in the early 1990s. The failure adequately to address the buffer zone problem then has led to the much more dangerous and costly confrontation we face today. Sometimes, the most valuable strategic analysis is that which forces us to confront uncomfortable truths about problems we would prefer to ignore.

The new Slavic states were never going to fit neatly into anyone's preferred world order. Recognizing this reality earlier might have prevented the current catastrophe.

We are in a period of significant historical change, and this book provides an intellectual photograph of some of the dynamics of this change, notably in the period of 2021 through 2025. The research was completed as of December 31, 2025.

INTRODUCTION

What began in February 2022 as Russia's invasion of Ukraine has fundamentally transformed into something far more consequential: a global contest between democratic and authoritarian powers that spans from the Korean Peninsula to the Atlantic.

As we reach the end of 2025, the conflict has evolved beyond Vladimir Putin's original territorial ambitions into a defining struggle that will determine the international order for the period ahead.

The war's expansion from a regional European conflict to a global coalition confrontation represents one of the most significant geopolitical realignments since the end of the Cold War. North Korean soldiers now fight alongside Russian forces in European trenches, while Japanese intelligence satellites provide crucial reconnaissance to Ukrainian defenders. Chinese economic support sustains Russia's war machine, even as South Korean weapons factories work overtime to supply NATO allies supporting Ukraine. This is no longer simply Putin's war. It has become a proxy conflict where Asian powers compete for influence on European battlefields.

The stakes have fundamentally shifted. What once appeared to be Putin's attempt to restore Russian imperial control over Ukraine has become a test case for whether authoritarian regimes can successfully use force to redraw international boundaries, with profound implications for Taiwan, the South China Sea, and democratic governance worldwide.

Putin is sealing Russia's fate for the decade ahead. As one strategic analyst wrote to me: "Connecting history and military strategies in the Ukraine War context context is crucial. Putin's chess playing strategies have sacrificed a decade or more of his country's future for what?"

THE AUTHORITARIAN AXIS: RUSSIA'S DEPENDENCE DEEPENS

The most dramatic transformation in the conflict has been North Korea's transition from arms supplier to active combatant. In a move unprecedented since the Korean War, Pyongyang has deployed troops to fight alongside Russian forces, marking Kim Jong Un's regime as Putin's most committed ally.[1]

Kim Jong Un's own words reveal the ideological dimensions of this commitment. In confirming the deployment through North Korea's state media, Kim declared that soldiers were sent to "annihilate and wipe out the Ukrainian neo-Nazi occupiers and liberate the Kursk area in cooperation with the Russian armed forces."[2] The North Korean leader has promised to build monuments in Pyongyang to honor those killed fighting Ukraine, treating these soldiers as heroes in a global struggle against Western influence.

This military partnership extends beyond personnel. North Korea has provided Russia with substantial quantities of artillery shells and ballistic missiles, deliveries that flagrantly violate multiple UN Security Council resolutions.[3] The mutual defense treaty signed between Putin and Kim in June 2024 has transformed what began as a transactional relationship into a formal military alliance.

The implications extend far beyond the immediate battlefield. North Korean forces are gaining invaluable experience in modern warfare, including exposure to drone technology, electronic warfare, and NATO-standard equipment, knowledge that could prove dangerous in any future Korean Peninsula conflict.

CHINA'S STRATEGIC SUPPORT AND CALCULATED DISTANCE

China's role remains more complex and carefully calibrated. While Beijing maintains what officials call a "no limits" strategic partnership with Russia, Chinese leaders have deliberately avoided the direct military involvement that characterizes North Korea's approach.

Instead, China provides crucial economic lifelines that enable Russia's war effort to continue. This includes energy purchases that help finance Moscow's military operations, dual-use technology transfers, and diplomatic cover in international forums. Chinese support has been essential in helping Russia circumvent Western sanctions and maintain access to global markets.

However, Beijing's support comes with important limitations. China has not provided lethal military aid directly, and Chinese officials have maintained studied silence about North Korean troop deployments. When pressed by international partners to rein in North Korea's involvement, Chinese Foreign Ministry spokespersons have insisted that Russia-North Korea relations are "a matter for themselves" as "independent sovereign states."[4]

This calculated approach reflects China's strategic calculations. While Xi Jinping's government wants to avoid Putin's "strategic defeat", understanding that Russian weakness could embolden Western pressure on China, Beijing also seeks to maintain relationships with European partners and avoid being drawn into direct confrontation with NATO.

The presence of Chinese and North Korean troops at Russia's Victory Day parade in May 2025 symbolized the alignment of these authoritarian powers, even as the relationships remain marked by underlying tensions and competing interests.

Although, ironically President Lula of Brazil was there too, clearly confusing BRIC solidarity with the Russian attack on the democratic order. This raises fundamental questions about Brazil's place in that order. This is a subject which I deal with in my 2026 book with Kenneth Maxwell which focuses on the relationships of Brazil and Australia with China. This book explores the key role of middle powers in shaping the new global system as the major powers struggle for position.

IRAN'S SUPPORTING ROLE

Iran completes the authoritarian axis through its provision of drone technology and technical expertise. Iranian-made Shahed drones have become a signature weapon of Russia's campaign against Ukrainian infrastructure, while Iranian technical advisors have helped establish production facilities within Russia itself.

This partnership has strengthened since the conflict began, with Iran seeing opportunity to field-test its weapons systems while supporting a fellow opponent of the Western-led international order. The collaboration has extended beyond military hardware to include intelligence sharing and coordination on sanctions evasion.

The impact of the Israeli air strikes on Iran clearly have had an effect on Iranian industry and its ability to support the Russians, but to what extent remains to be seen.

THE DEMOCRATIC
COALITION RESPONSE

The most significant institutional response to authoritarian align-ment has been the emergence of the multi-country "Coalition of the Willing" spearheaded by the United Kingdom and France. This diplomatic structure represents Europe's most ambitious attempt to organize sustained support for Ukraine while preparing for post-conflict security arrangements.

The coalition's formation reflects European recognition that tradi-tional institutional frameworks both NATO and the EU face limita-tions in responding to the crisis. NATO consensus remains elusive on direct involvement, while EU decision-making has been hampered by Hungarian and Slovakian opposition to military support measures.

The coalition has committed to provide at least €40 billion in military support to Ukraine throughout 2025, matching NATO's previous commitments.[5] More significantly, it has developed detailed operational plans for a "Multinational Force Ukraine" ready to deploy once hostilities cease. This force would help secure Ukraine's airspace and territorial waters while supporting the regeneration of Ukrainian armed forces.

The coalition's membership reveals the breadth of European commitment. It includes most EU members, with notable exceptions being Malta and Austria (due to neutrality policies) and Hungary and Slovakia (due to policy disagreements with military support). Significantly, traditionally neutral Ireland has joined, with its leadership declaring readiness to "do whatever we possibly can to help."

Beyond Europe, the coalition includes key partners from other regions: Canada, Australia, New Zealand, Turkey, and crucially, representatives from Japan and South Korea. This geographic diversity underscores how the conflict has transcended regional boundaries.

NATO'S ADAPTATION AND EXPANSION

While the Coalition of the Willing operates outside formal NATO structures, the alliance itself has adapted significantly to support Ukraine. The NATO Security Assistance and Training for Ukraine (NSATU) program, operational since January 2025 with 700 personnel, coordinates capability coalitions across specific military domains, air defense, artillery, logistics, and cyber warfare.[6]

NATO has also welcomed unprecedented engagement from Indo-Pacific partners. Japan opened a dedicated diplomatic mission to NATO, while South Korea has doubled its contributions to NATO's Ukraine Trust Fund and signed agreements for mutual recognition of military standards.

These developments represent NATO's recognition that security challenges are increasingly global rather than regional. The alliance's traditional focus on Euro-Atlantic security has expanded to acknowledge the interconnected nature of authoritarian challenges from Eastern Europe to East Asia.

Murielle Delaporte and I already anticipated this shift when we published our book in 2020 on *The Return of Direct Defense in Europe: Meeting the 21st Century Authoritarian Challenge*. We deliberately expanded the discussion beyond the question of Russia and its challenge to Europe to the broader consideration of the challenge of the global multi-polar authoritarian order to Europe.

INDIVIDUAL EUROPEAN COMMITMENTS

Beyond collective frameworks, individual European nations have made substantial independent commitments. Germany, despite initial reluctance, has become a major provider of advanced weapons systems including Taurus missiles and sophisticated air defense systems. The Nordic countries, Denmark, Sweden, Norway, and Finland, have provided comprehensive support packages that include both military aid and long-term security commitments.

Poland has emerged as a crucial logistics hub and staging area for Western support, while also serving as a bridge between European and American defense industries. The Baltic states, Estonia, Latvia, and Lithuania, have provided some of the highest levels of support relative to their GDP, reflecting their understanding that Ukrainian victory is essential for their own security.

This pattern of European commitment reflects a fundamental shift in strategic thinking. German Chancellor Friedrich Merz warned that if Ukraine capitulates, "the day after tomorrow it will be the next country's turn, and then the day after that it will be our turn. That's not an option." This was stated in an interview with the German channel ZDF in August 2025, a view widely reflected in contemporary European strategic discourse. Multiple think-tank reports and policy analyses echo the idea that failing to support Ukraine could embolden further Russian aggression across Europe.[7]

ASIAN POWERS ENTER
THE EUROPEAN WAR

Japan's involvement in the Ukraine conflict represents one of the most significant shifts in its post-World War II foreign policy. Prime Minister Kishida's March 2023 visit to Kyiv, making him the first Japanese leader to visit a conflict zone since 1945, symbolized Tokyo's transformation from pacifist observer to active participant in global security affairs.

Japan has emerged as one of Ukraine's largest bilateral donors, providing approximately $12 billion in aid including both humanitarian assistance and military equipment. This support includes non-lethal military supplies such as reconnaissance drones, protective equipment, and vehicles, carefully calibrated to remain within Japan's constitutional constraints while maximizing impact.[8]

More significantly, Japan has provided crucial intelligence support through its advanced satellite reconnaissance capabilities. Japanese satellite data has proven invaluable for Ukrainian forces in tracking Russian troop movements and planning operations. This intelligence sharing represents a level of operational cooperation that would have been unthinkable just a few years ago.

Japan's motivation stems from its understanding that "Ukraine today may be East Asia tomorrow."[9] The precedent of successful territorial conquest through military force would directly threaten Japanese security interests regarding Taiwan and disputed territories in the East China Sea.

The conflict has also accelerated Japan's defense transformation. Tokyo has committed to increasing defense spending to 2% of GDP by 2027, a historic 60% increase that will give Japan the world's third-largest defense budget. [10]

South Korea's involvement has evolved from initial reluctance to increasingly robust support. Under President Yoon Suk-yeol, Seoul has moved from providing humanitarian aid to considering direct military assistance, particularly following confirmation of North Korean troop deployments in Russia.

South Korea has become NATO's most engaged Indo-Pacific partner, attending three consecutive NATO summits and establishing unprecedented cooperation agreements. Seoul has doubled its contributions to NATO's Ukraine Trust Fund, committing $24 million for 2025, while signing agreements for mutual recognition of military airworthiness standards.[11]

North Korea's direct involvement in the war (with approximately 10,000-12,000 troops deployed to Russia) led South Korean President Yoon Suk-yeol to announce that Seoul would reconsider its long-standing policy against providing lethal weapons to Ukraine. He stated that while South Korea had "a principle of not directly supplying lethal weapons" to combatants, the country "can be more flexible and review the policy depending on North Korea's military activities."

This represented a significant shift from South Korea's previous position, as the North Korean troop deployment was seen as directly affecting South Korean security interests in ways that humanitarian aid to Ukraine had not.[12]

Australia has provided limited specialized support focused on advanced military technologies. This support reflects Australia's recognition that successful resistance to authoritarian aggression in Europe strengthens deterrence in the Indo-Pacific.

Australian involvement has been coordinated closely with the United States and other Five Eyes intelligence partners, ensuring that technological transfers to Ukraine support broader Western strategic objectives. Australia's participation in the Coalition of the Willing and NATO partnership programs demonstrates Canberra's commitment to what officials call "integrated deterrence" across multiple theaters.

MILITARY DYNAMICS AND
BATTLEFIELD REALITY

The human cost of Russia's expanded war effort has reached staggering proportions. Intelligence assessments indicate that Russian military casualties are approaching the one million mark as summer 2025 progressed, a devastating toll that raises fundamental questions about the sustainability of Putin's campaign.[13]

Equipment losses have been equally severe. Since January 2024, Russia has lost approximately 1,149 armored fighting vehicles, 3,098 infantry fighting vehicles, 300 self-propelled artillery pieces, and 1,865 tanks. Perhaps most significantly, Russian equipment losses have consistently exceeded Ukrainian losses at ratios varying between 2:1 and 5:1 in Ukraine's favor.[14]

These losses reflect the fundamental challenges facing Russian forces. Despite numerical advantages and support from North Korean troops, Russian advances have slowed dramatically. Along the crucial Donetsk front, Russian forces have averaged just 135 meters of daily advance, a pace that makes strategic breakthrough virtually impossible.

The inefficiency of Russian operations has become particularly apparent around key targets like Pokrovsk, where months of assault have yielded minimal territorial gains at enormous cost. The arrival of North Korean reinforcements has provided additional manpower but has not fundamentally altered the battlefield dynamics.[15]

Ukrainian forces have demonstrated remarkable adaptability in countering Russian offensives supplemented by North Korean troops. Initial reports suggested that North Korean forces, lacking experience with modern drone warfare and armored vehicle operations, suffered heavy early casualties. However, these units have reportedly adapted quickly to battlefield conditions.[16]

The integration of Western training and equipment has proven crucial to Ukrainian defensive success. NATO-standard training programs, implemented through NSATU and bilateral agreements, have enabled Ukrainian forces to effectively employ increasingly sophisticated weapon systems while maintaining operational coordination with Western intelligence assets.

Ukrainian access to Japanese satellite intelligence, European drone technology, and American precision weapons has created a comprehensive defensive capability that continues to frustrate Russian offensive operations despite the addition of North Korean reinforcements.

Russian summer offensive operations led to territorial gains which were modest despite intensified fighting. Russian advances around towns like Pokrovsk and Kostiantynivka have measured in kilometers rather than the tens of kilometers needed for strategic impact. The slow pace of advance, combined with heavy casualties, suggests that Russia's enhanced coalition with North Korea has increased capability but not fundamentally altered battlefield dynamics.

Ukrainian defensive operations in Kursk region, where North Korean troops have been most heavily engaged, continue despite Russian claims of complete recapture. The ability of Ukrainian forces to maintain defensive positions in Russian territory, even against combined Russian-North Korean assault, demonstrates the continued effectiveness of Western-supported Ukrainian military capabilities.

THE AMERICAN QUESTION

The return of Donald Trump to the presidency has introduced significant uncertainty into coalition dynamics. Trump's criticism of continued Ukrainian support and his stated desire to end the conflict "in a day" have raised questions about long-term American commitment.

However, the involvement of senior American officials, including General Keith Kellogg, in Coalition of the Willing meetings suggests continued American engagement even under changed political leadership.[17] The participation of Republican senators alongside

European partners indicates that American support retains bipartisan elements despite presidential ambivalence.

The European response to potential American disengagement has been to accelerate independent capabilities while maintaining coordination with Washington. This approach reflects European determination to maintain support for Ukraine regardless of American political changes.

Despite presidential uncertainty, American military and intelligence cooperation with both Ukraine and coalition partners has continued at operational levels. The integration of American capabilities with Japanese intelligence, European logistics, and Ukrainian operations demonstrates institutional momentum that transcends political changes.

Congressional support for coalition approaches has remained relatively stable, with bipartisan recognition that successful authoritarian aggression in Europe would create dangerous precedents for American interests globally.

CONCLUSION

The outcome of the Ukraine conflict has become inseparable from broader questions of international order. Successful Russian territorial conquest, even with significant costs, would establish a dangerous precedent for other revisionist powers. Chinese strategists are undoubtedly studying the conflict for lessons applicable to potential Taiwan scenarios.

Conversely, Russian failure would demonstrate the effectiveness of coordinated democratic resistance to authoritarian aggression. The coalition model developed in response to Russian invasion could serve as a template for addressing future challenges in other regions.

The conflict has accelerated the evolution of traditional alliance structures. NATO's incorporation of Indo-Pacific partners reflects recognition that security challenges are increasingly global. The Coalition of the Willing model provides flexibility that formal alliance structures often lack.

These developments suggest a future international system characterized by overlapping coalitions rather than rigid blocs. Democratic nations are developing multiple frameworks for cooperation that can be activated based on specific challenges rather than relying solely on Cold War-era institutions.

And these democratic coalitions are no longer going to be led automatically by the United States. American dominance is over whether you are looking at the Biden or Trump Administrations. The democratic middle powers need to seriously think through their relationships with the authoritarian powers for their policies are fueling the rise of the authoritarian axis, for their is no longer going to be an American policeman for the liberal world order.

The conflict has also intensified economic and technological competition between democratic and authoritarian coalitions. Western sanctions have forced Russia toward greater dependence on Chinese and North Korean support, while accelerating European efforts to reduce strategic dependencies.

The race to develop and deploy advanced military technologies, from drone systems to missile defense, has become a crucial element of the broader competition. Japanese satellite capabilities, South Korean weapons production, and European defense innovation have all proven essential to supporting Ukrainian resistance.

The transformation of the Russia-Ukraine conflict from regional war to global coalition confrontation represents a fundamental shift in how international conflicts develop and are contested. The emergence of cross-regional partnerships, North Korean soldiers fighting in Europe, Japanese satellites supporting Ukrainian operations, European weapons flowing through South Korean production lines, demonstrates the increasingly global nature of strategic competition.

This model may define future conflicts. The coalition structures developed in response to Russian aggression provide templates for addressing other challenges, while the precedents established will influence authoritarian calculations for the period ahead. The successful integration of Asian democratic powers into European security challenges suggests possibilities for coordinated responses to future crises.

Putin's invasion of Ukraine has indeed become something far larger than its original scope. It has become a test of whether democratic coalitions can effectively resist authoritarian aggression in an era of global strategic competition. The outcome will determine not just Ukraine's future, but the broader structure of international order in the 21st century.

The war's expansion into a global contest between democratic and authoritarian coalitions represents both opportunity and danger. Success in supporting Ukrainian resistance while building effective coalition structures could establish robust precedents for addressing future challenges. Failure could encourage further authoritarian aggression and undermine the international system that has underpinned global stability since 1945.

What began as Putin's attempt to restore Russian imperial control has become a defining struggle for the future of international order. The coalitions now engaged in this struggle, democratic and authoritarian alike, understand that the outcome will shape global politics for decades to come.

The transformation from regional war to global coalition conflict is now complete. The question remaining is which model of international order will emerge from this crucible: one based on democratic cooperation and respect for sovereignty, or one where authoritarian powers can successfully use force to reshape international boundaries.

The answer will be determined not just by events on Ukrainian battlefields, but by the strength and persistence of the democratic coalitions now committed to Ukraine's survival and success. And recognizing, of course, that Ukraine is not a democracy itself, but rather the location where the conflict affecting the fate of the democratic global order is being played out.

SECTION I
BACKGROUND PERSPECTIVES

CHAPTER 1

THE PUTIN DYNAMIC: THE PRELUDE TO THE WAR

Vladimir Putin's rise from a relatively obscure KGB officer in Dresden to the architect of Russia's post-Soviet revival and decline represents one of the most consequential political transformations of the early twenty-first century.

Over three decades, Putin forged a narrative that filled the void left by the Soviet Union's collapse, created a centralized authoritarian state, and positioned Russia to challenge the Western-dominated international order.

This chapter examines the Putin dynamic from the chaos of the 1990s through his consolidation of power in the 2000s and 2010s, exploring how he crafted a distinctly Russian form of twenty-first-century authoritarianism that continues to shape global politics.

THE FORMATIVE YEARS: DRESDEN AND THE SOVIET COLLAPSE (1980-1991)

Vladimir Putin's worldview was fundamentally shaped by his experiences in the final decade of the Soviet Union. As a KGB officer stationed in Dresden, East Germany, during the 1980s, Putin witnessed firsthand the superpower confrontation between East and

West. This period coincided with Ronald Reagan's aggressive campaign to curtail Soviet power, culminating in the Euro-missile crisis that made Europe a central theater of Cold War conflict.

In Dresden, Putin's responsibilities extended beyond routine intelligence work. He developed fluency in German and cultivated networks within the East German STASI that would prove valuable in his later political career. More importantly, he absorbed KGB techniques for understanding and influencing Western Europe, with particular emphasis on exploiting divisions between European nations and the United States. These lessons in leveraging Western disunity would become a cornerstone of his later foreign policy.

The fall of the Berlin Wall in 1989 delivered a profound shock to Putin and his generation. As Angus Roxburgh observed, when Boris Yeltsin appointed Putin prime minister in August 1999, few Russians knew much about this "mousy, shy and awkward" figure.[1] But his experiences watching the Soviet empire crumble in Dresden had forged a ruthless determination that would soon become his defining characteristic.

The dissolution of the Soviet Union on December 26, 1991, when the Supreme Soviet voted the USSR out of existence, represented far more than a geopolitical shift. It triggered the collapse of an entire interconnected system of political power and imperial economy spanning eleven time zones. For Putin's generation, this was their "time of troubles" or a descent into chaos that would profoundly influence their understanding of state power and national destiny.

THE WILDERNESS YEARS: RUSSIA IN THE 1990S

When I visited Russia in the early 1990s, it resembled entering a frontier country with its own version of the Wild West, where the distinctions between cowboys, sheriffs, and bad guys remained perpetually unclear.

The decade following Soviet collapse subjected ordinary Russians to extraordinary hardship. The centralized Soviet economic system's implosion left pensioners destitute, infrastructure crumbling, and organized crime flourishing in the regulatory vacuum.

During this period, Putin found his footing in the reform movement under Anatoly Sobchak, the democratic mayor of St. Petersburg (newly restored from its Soviet name, Leningrad). Sobchak, who emerged as a leading democratic voice alongside Boris Yeltsin in the late 1980s, gave Putin his start in public life. Working for Sobchak provided Putin with crucial experience in navigating the chaotic post-Soviet political landscape and building a network of loyal associates who would later form the core of his presidential administration.[2]

When Sobchak lost his reelection bid in 1996 and fled to Paris facing corruption charges, Putin's loyalty to his mentor demonstrated a personal code that would characterize his leadership style. Despite being out of work, Putin maintained his Petersburg networks, relationships that would prove invaluable as he climbed the ranks of Boris Yeltsin's administration.

Putin's trajectory through the Yeltsin administration revealed his ability to navigate bureaucratic infighting and position himself as an indispensable operative. In March 1997, he became head of the Main Control Directorate and deputy chief of staff in the presidential administration. This appointment provided him with comprehensive knowledge of Russia's regional governance and contacts with the prosecutor's office and Federal Security Service (FSB), the KGB's successor organization.

By May 1998, Yeltsin appointed Putin to head the FSB itself, placing him at the center of Russia's security apparatus. This rapid advancement through three major positions in less than two years positioned Putin uniquely to understand the levers of state power while the Yeltsin administration entered its death throes.

WHAT THE WEST GETS
WRONG ABOUT PUTIN

My friend Harald Malmgren wrote an essay on his meeting with Putin in the early 1990's which provides an interesting set of insights about the dynamic younger Putin before he became the aging Tsar of today.

This is what Hal wrote:

July 7, 2022

 In 1999, Vladimir Putin suddenly sprang from bureaucratic obscurity to the office of Prime Minister. When, a few months later, Yeltsin unexpectedly resigned and Putin was voted in as President, governments around the world were taken by surprise yet again.

 How could this unknown figure have amassed national voter support with so little media attention?

 I had first met Putin in 1992 and was not surprised by his rapid domination of the new Russia. We were introduced by Yevgeny Primakov, widely known as "Russia's Kissinger" whom I had met in Moscow multiple times during the Cold War years when I advised Presidents Kennedy, Johnson, Nixon and Ford.

 Primakov was a no-nonsense thinker and writer. He was also a special emissary for the Kremlin in conducting secret discussions with national leaders around the world.

 When Yeltsin tasked his advisor Anatoly Sobchak with identifying and recruiting Russia's best and brightest, Putin, then a local politician in his hometown of St Petersburg, was top of his list — so Primakov took Putin under his wing to tutor him in global power and security issues. Eventually, Primakov introduced Kissinger to Putin, and they became close.

 That both Primakov and Kissinger took time to coach Putin on geopolitics and geo-security was a clear demonstration that they saw in him the characteristics of a powerful leader. It also showed Putin's capacity for listening to lengthy lessons on geopolitics — as I was soon to learn.

 In 1992, I received a call from a meeting organizer at the CSIS think tank inviting me to join a U.S.-Russia St Petersburg Commission to be chaired by Kissinger and Sobchak.

 The purpose would be to help the new Russian leadership in opening channels of business and banking with the West. Most of the Western members would be CEOs of major U.S. and European companies, as well as key officials of the new Russian government. I would attend as an expert.

I was told that a "Mr Primakov" had personally asked if I could make time to participate. I could hardly refuse such a request, and I was intensely curious about the emerging Russian leadership, especially about Putin.

Arriving at the first meeting, I saw several people gathered around Kissinger and a man I was told was Putin. An official identified himself to me and said he had been asked by Primakov to introduce me to Putin. He interrupted the conversation with Kissinger to announce my arrival; Putin warmly responded that he was looking forward to chatting with me about how I see the world from inside Washington.

We spoke on several occasions between meetings, and he arranged to sit next to me at a dinner, accompanied by his interpreter. At that dinner, he asked me: "What is the single most important obstacle between your Western businessmen and my fellow Russians in starting up business connections?"

Off the top of my head, I responded: "The absence of legally defined property rights — without those there is no basis for resolving disputes."

He responded:

"Ah yes, in your system a dispute between businesses is resolved by attorneys paid by the hour representing each side, sometimes taking the dispute to the courts which normally takes months and accumulation of hourly attorney fees.

"In Russia, disputes are usually resolved by common sense. If a dispute is about very significant money or property, then the two sides would typically send representatives to a dinner. Everyone attending arriving would be armed. Facing the possibility of a bloody, fatal outcome both sides always find a mutually agreeable solution. Fear provides the catalyst for common sense."

He used his argument in the context of disputes between sovereign nations. Solutions often require an element of fear of disproportionate responses if no deal is struck. The idea of forcing adversaries to face horrific alternatives seemed to excite him. In essence, he was describing to me the current Ukraine impasse between the U.S. and Russia.

Putin knows Russia cannot afford a prolonged ground war with Ukraine. He also can see Biden is facing crucial midterm elections with a domestic congressional impasse and cannot afford a major foreign crisis distraction. The two sides have no choice but to strike a deal.

On a different occasion, Putin asked me how decisions are really made in Washington, with its complex division of Presidential and Congressional powers. He said Kissinger could explain the broad parameters of a Presidential policy decision, but could not clarify how political consensus was achieved between the House, Senate, and the Executive Branch.

It was evident he had been given a deep intelligence brief on my career. He said Kissinger enjoys the public theatre of powerful people meeting in elaborate dinners or meetings with many aides ready to guide them. And he told me he had been informed that I preferred backroom meetings to shape consensus and provide room for negotiating details.

I tried to explain the elaborate process of balancing the interests of the many players in Washington, including Congress, the major agencies, and the intricate business arrangements that might be affected by any decision.

I told him of my first personal meeting with Nixon, who had said he was impressed that I had strong personal support from leaders of both major parties. However, he added, this raised worries among his staff in the White House — so he really needed to know whether I was a Republican or a Democrat. To which I replied: "Yes."

When Nixon asked what that meant, I explained that I was not a partisan warrior, but rather a problem solver. To get a solution I would always be ready to work with key players of both parties depending upon the specific problem. This seemed to amuse Putin.

The impression of Putin that I was left with was of a man who was more intelligent than most of the politicians I had met in Washington and in other capitals around the world. I was reminded of my childhood: I grew up in a predominantly Sicilian neighborhood, with a mafia maintaining order. No disorganized crime allowed.

Putin did seem to have the instincts of a Sicilian mafia boss: quick to reward but quick to pose mortal risk in the event of non-conformity to the family rules.

Looking back to those times of growing disarray in Russia's leadership, I can recall the prolonged, multi-year paralysis of the Brezhnev Presidency, which was followed by the brief Presidencies of Andropov and Chernenko.

Gorbachev was not strong enough to impose his will. Yeltsin had good ideas but was easily distracted and lacked follow through. Russia was in urgent need of a strong leader — and so Putin stepped in.

As for how Putin sees himself, he did bring up several times his admiration for Peter the Great, so much so I was convinced he sees himself as his incarnation.

I have not been a guest of the Kremlin since 1988, but I am told Putin had portraits of Peter the Great hung in several important meeting rooms there — rather than portraits of himself, as would be more customary.

What this means for Biden, Nato and Ukraine is slowly becoming clear. There is more to Putin than meets the eye.

This article was first published UnHerd.com.[3]

It is republished with the author's permission.

THE CRUCIBLE OF CHECHNYA AND KOSOVO (1999)

Two international crises in 1999 would profoundly shape Putin's emergence as a national figure and define his approach to state power. The renewed turmoil in Chechnya coincided with NATO's intervention in Kosovo, creating a perfect storm that inflamed Russian nationalist sentiment and provided Putin with opportunities to demonstrate the ruthless leadership that would become his trademark.

As Steven Lee Myers noted, the prospect of NATO military intervention to protect Kosovo infuriated Russia in ways American and European leaders failed to appreciate. Russia shared Slavic roots, religion, and culture with Serbia, but Russian concerns extended beyond ethnic solidarity. The conflict inflamed Russia's wounded pride over its deflated post-Soviet status and inability to shape world events. Many Russians, not prone to paranoia, could easily imagine a future NATO campaign on behalf of Chechnya's independence movement.[4]

Putin's response to the Chechen situation revealed the character that would define his rule. His infamous threat to wipe out terrorists "even if they're in the shithouse" signaled a departure from the hesitant, alcoholic leadership of Yeltsin's final years.[5] Within weeks, Putin had launched a devastating war against Chechen separatists that would leave tens of thousands of civilians dead, establishing his credentials as a leader willing to use overwhelming force to maintain state integrity.

The Kosovo crisis reinforced Putin's conviction that the West, particularly the United States, operated with hypocritical exceptionalism in international affairs. This perception would fuel his later challenges to the Western-dominated order and provide rhetorical justification for Russian actions in Georgia and Ukraine.

THE MILLENNIUM MESSAGE: PUTIN'S VISION FOR RUSSIA (1999-2000)

On December 30, 1999, in his famous turn-of-the-millennium address, Putin articulated a vision for Russia that would guide his entire tenure. This speech deserves careful attention as it laid bare his understanding of Russian political culture and his plans for reconstruction:

> *It will not happen soon, if it ever happens at all, that Russia will become the second edition of, say, the U.S. or Britain in which liberal values have deep historic traditions. Our state and its institutes and structures have always played an exceptionally important role in the life of the country and its people. For Russians a strong state is not an anomaly which should be got rid of. Quite the contrary, they see it as a source and guarantor of order and the initiator and main driving force of any change.[6]*

This passage reveals Putin's fundamental rejection of Western-style liberal democracy as incompatible with Russian political culture. He positioned himself not as an opponent of all democratic forms, but as an advocate for a distinctly Russian version of governance that emphasized state power as the primary guarantor of order and prosperity.

Putin continued:

Modern Russian society does not identify a strong and effective state with a totalitarian state. We have come to value the benefits of democracy, a law-based state, and personal and political freedom. At the same time, people are alarmed by the obvious weakening of state power. The public looks forward to the restoration of the guiding and regulating role of the state to a degree which is necessary, proceeding from the traditions and present state of the country.

This rhetorical sleight of hand, acknowledging democratic values while prioritizing state power, would characterize Putin's approach throughout his rule. He presented himself as offering a middle path between totalitarianism and the chaos of the Yeltsin years, though in practice his regime would steadily erode the democratic gains of the 1990s.

When Yeltsin surprised the world by resigning and designating Putin as acting president on December 31, 1999, this speech provided the ideological framework for the transformation that would follow. Putin won the presidential election in March 2000 with a clear mandate to restore order and state authority.

REBUILDING THE STATE (2000-2007)

Putin's first phase in power focused on reconstructing centralized state authority and positioning himself to control Russia's vast energy resources. His strategy drew on work begun during his Petersburg years, when he participated in informal discussions about natural resources under Vladimir Litvinenko at the Mining Institute, where Putin had defended his thesis.

The dramatic rise in global energy prices during the 2000s provided Putin with the financial resources to fund pension restoration, rebuild infrastructure, and generally improve living standards for ordinary Russians. This economic revival proved crucial for establishing his legitimacy and popular support. Russians who had suffered through the 1990s chaos welcomed the stability and modest prosperity Putin delivered, even as he systematically dismantled checks on executive power.

During this period, Putin pursued a complex relationship with the West. Following the September 11, 2001 terrorist attacks, he became the first foreign leader to call President George W. Bush, offering support and cooperation. Putin overruled his military advisors to permit the United States to use Russian transit points for military operations in Afghanistan. In return, he expected Western acceptance of Russian actions in Chechnya, framed as part of the global war on terrorism, and recognition of Russia as a great power deserving a seat at the table in shaping international security. In fact, Putin thinks very much in terms of classic great power terms, rather than a globalization narrative.

However, neither the Clinton nor Bush administrations proved eager to embrace Putin's vision of Russia's role. From Putin's perspective, the West, particularly the United States, exploited Russia's weakness during the 1990s to expand NATO eastward, incorporating former Soviet satellites and Baltic states. While Western leaders characterized this expansion as the free choice of sovereign nations, Putin viewed it as a deliberate strategy to encircle and weaken Russia while preventing its return to great power status.

The expansion of the European Union, led by Germany, compounded Putin's concerns. Rather than viewing former communist states as newly sovereign nations making independent choices, Putin saw them as within Russia's legitimate sphere of influence or territories where Russia, as a great power, possessed veto rights over major strategic decisions.

THE MUNICH DECLARATION: DRAWING LINES (2007)

Putin's February 2007 speech at the Munich Security Conference marked a turning point in Russia's relationship with the West. After seven years of attempting to position Russia as a partner deserving equal status with Western powers, Putin delivered a blunt assessment of international relations that shocked his audience:

> *Today we are witnessing an almost uncontained hyper use of force, military force, in international relations, force that is plunging the world into an abyss of permanent conflicts. As a result, we do not have sufficient strength to find a comprehensive solution to any one of these conflicts. We are seeing a greater and greater disdain for the basic principles of international law. And independent legal norms are, as a matter of fact, coming increasingly closer to one state's legal system. One state and, of course, first and foremost the United States, has overstepped its national borders in every way.* [7]

Putin's critique extended to NATO expansion, which he characterized as a "serious provocation that reduces the level of mutual trust." He pointedly asked: "Against whom is this expansion intended?" He reminded his audience of assurances given after the Warsaw Pact's dissolution that NATO would not expand eastward, though Western leaders contested whether such guarantees were ever formally made.

The speech revealed Putin's growing frustration with what he perceived as Western hypocrisy and disregard for Russian interests. In his view, he had attempted to work within the Western-dominated order, but now signaled Russia's intention to challenge that order and defend its perceived sphere of influence through whatever means necessary.

THE MEDVEDEV INTERLUDE (2008-2012)

Constitutionally barred from a third consecutive term in 2008, Putin orchestrated an arrangement whereby his protégé Dmitry Medvedev became president while Putin assumed the role of prime minister. This period revealed the flexibility of Putin's power structure whereby he maintained control over foreign and defense policy while nominally serving in a subordinate position.

The 2008 Russia-Georgia War demonstrated Putin's willingness to use military force to enforce Russian interests in the "near abroad." When Georgia, seeking closer ties with NATO and the European Union, clashed with Russian-backed separatists in South Ossetia and Abkhazia, Russian forces intervened decisively, routing Georgian forces and seizing Georgian territory. From Putin's perspective, the Bush administration's limited response validated his strategy of using military force to halt NATO's eastward expansion.[8]

This period also saw the Obama administration's attempted "reset" of U.S.-Russia relations, an initiative Putin viewed as an opportunity to leverage while continuing his internal consolidation of power. The reset ultimately failed to reconcile fundamentally incompatible visions of international order, but it provided Putin breathing room to deepen his control over Russia's eleven time zones through his "vertical power" system.

RETURN AND
CONFRONTATION (2012-2014)

Putin's return to the presidency in 2012 marked the beginning of his most assertive phase. The protests that greeted his return, sparked by allegations of electoral fraud, reinforced his conviction that Western-style democracy posed a threat to Russian stability. He responded by tightening controls on civil society, media, and political opposition while ramping up nationalist rhetoric.

Ukraine's Euromaidan protests in 2013-2014, which toppled the pro-Russian president Viktor Yanukovych, represented Putin's ultimate red line. The prospect of Ukraine joining the European Union and potentially NATO was intolerable to Putin's vision of Russia's legitimate sphere of influence. Putin had grudgingly accepted NATO expansion into the Baltic states, but influential American and European officials now openly were advocating including Ukraine into NATO, which for Putin was a step too far.

The seizure of Crimea in March 2014 marked a watershed moment in post-Cold War European history. Putin justified the annexation by claiming to protect ethnic Russians and invoking Kosovo as precedent, though the comparison was deeply flawed. As Shaun Walker noted in *The Long Hangover*, it was absurd to compare Kosovo, which became independent after sustained ethnic cleansing, with Crimea, which was seized by its larger neighbor based on theoretical and greatly exaggerated threats.[9]

Nevertheless, Putin's gambit succeeded domestically for the annexation proved wildly popular with Russians, validating his narrative about defending Russian interests against Western encroachment. Putin had complained about American exceptionalism in Munich back in 2007. Now he had done something about it. The Crimea operation also addressed specific strategic concerns about Ukraine tilting decisively toward the West and potentially expelling Russia's Black Sea Fleet from its Crimean bases.

THE PUTIN NARRATIVE: NATION-BUILDING THROUGH MEMORY AND MYTH

Throughout his tenure, Putin systematically constructed a narrative about Russia and its place in the world that resonated with citizens who had suffered through the 1990s chaos. As Shaun Walker argued in *The Long Hangover*, Putin's mission was to fill the void left by the 1991 collapse and forge a new sense of nation and purpose. Rather than simply building Soviet Union 2.0, Putin selectively leveraged aspects of Russian and Soviet history to create something new.

Central to this narrative was the fetishization of "stabilnost" (stability), which meant that revolution and state collapse were inherently bad, whether in 1917, 1991, or in any hypothetical future Russia.

Putin also weaponized World War II memory, transforming the Great Patriotic War into a central pillar of Russian national identity. Annual Victory Day celebrations became increasingly militarized spectacles, with the narrative emphasizing Soviet heroism while downplaying or ignoring Stalin's catastrophic mistakes that contributed to the war's outbreak and initial Soviet defeats. The simplified narrative served Putin's purposes: Russia must unite against foreign threats, just as it did against Nazi Germany.

As Shaun Walker assessed it, Putin had largely succeeded in creating a sense of nation and rallying Russia around a patriotic idea.[10] However, instead of transcending the trauma of Soviet collapse, his government exploited it, using fear of political unrest to quash opposition, equating patriotism with support for Putin, and employing a simplified war narrative to suggest Russia faced existential foreign threats requiring unity under strong leadership.

BREAKING OUT: THE SYRIAN INTERVENTION (2015-2020)

Facing Western sanctions and diplomatic isolation following the Ukraine crisis, Putin found new opportunities in Syria's civil war. The Russian military intervention beginning in September 2015 represented a dramatic departure from previous post-Soviet military operations. As Dmitri Trenin observed, this was Russia's biggest combat deployment abroad since Afghanistan, but it represented a fundamentally different kind of warfare.[11]

The Syrian operation was expeditionary, with no common border with Russia, predominantly aerial, with minimal ground forces, coalition-based, requiring coordination with Syrian and Iranian forces, deliberately limited, and closely tied to diplomatic processes. These characteristics demonstrated Russia's evolution in employing military power as a tool of foreign policy.

The intervention achieved multiple objectives simultaneously. It preserved the Assad regime, a long-time Soviet ally, while establishing permanent Russian air and naval bases in the Eastern Mediterranean. It provided combat experience for testing new weapons systems and operational concepts. Most importantly, it positioned Russia as an indispensable diplomatic player in the Middle East, breaking the post-Crimea stalemate in Europe through strategic maneuvering in a different theater.

As Trenin underscored, Moscow emerged from Syrian engagement as the regional player with the most connections, maintaining close contacts with virtually all Middle Eastern leaders despite their mutual antagonisms: Turkey and the Kurds, Iran and Saudi Arabia, Iran and Israel. This ability to navigate complex regional divides while promoting Russian interests marked a sophisticated approach to twenty-first-century power projection.

Putin chose the Middle East as the venue for demonstrating that Russia had returned to global great power status. The combination of clear objectives, strong political will, area expertise, resourceful diplomacy, capable military forces, and ability to coordinate with diverse partners in a complex region enabled Russia to project power onto the top level of global politics which was exactly what Putin aimed to achieve.

THE CODE OF PUTINISM: IDEOLOGY AND PRACTICE

Brian Taylor, in *The Code of Putinism*, provided an insightful framework for understanding Putin's ideological system. At its core lies the notion that Russia must be a strong state and great power, deeply suspicious of the United States and the global system it promotes. Feelings of humiliation and resentment following the Soviet collapse fuel this drive to restore Russian power.[12]

Putin's Russia does not seek global hegemony but aims to upend the Western-dominated order promoting democracy and human rights. The code embraces conservative and illiberal values emphasizing tradition, hierarchy, and order. Putin, increasingly skeptical of reaching accommodation with the West, sought to undermine the image of the West and its liberal values both domestically and internationally. His narrative positioned Russia as a defender of traditional values against Western decadence and moral decay, though this message competed awkwardly with his support for authoritarian regimes like Iran and China that held fundamentally different value systems.

Taylor emphasized that Putin's Russia wants great power diplo-

macy as practiced in nineteenth and twentieth-century Europe or a
concert of nations approach that most of the international system,
particularly European states and Russia's neighbors, no longer
accepts as viable. The countries where Russia used military force
without invitation, Moldova, Georgia, Ukraine, are precisely those
most resistant to remaining in Russia's orbit, validating the observation that the harder Russia squeezes its neighbors, the more they seek
refuge in Euro-Atlantic institutions.

THE LIMITS OF PUTINISM

Despite Putin's successes in rebuilding Russian state power and challenging Western dominance, his system contains inherent contradictions and limitations. Tony Wood, in *Russia Without Putin*, argued
that the sweep of history since Soviet collapse represents a continuous development of Russian-style capitalism and authoritarianism,
not a radical break between Yeltsin and Putin eras. Both phases
contributed to establishing the current system's foundations.

Wood highlighted that for much of the post-Soviet era, Russia's
elite, including Putin, pursued integration or alliance with the West.
Over time, as this proved illusory, Russia's elite gradually abandoned
the fantasy, replacing dreams of integration with more strident
defense of Russian interests. However, this shift created new vulnerabilities.[13]

Russia's relative weakness compared to potential competitors
constrains its options. If China becomes the twenty-first century's
dominant superpower, Russia shares an extensive land border with
that power and would Russia become China's Mexico, economically
integrated with and strategically subordinated to its giant neighbor?

THE MISSING ELEMENT:
STALIN'S LEGACY

A critical blind spot in Putin's historical narrative concerns authoritarian leadership's potential for catastrophic error. Putin's selective

memory of World War II celebrates Soviet heroism while obscuring Stalin's role in enabling the conflict. Stalin blocked German communists and socialists from cooperating against the Nazis, facilitating Hitler's rise. The Molotov-Ribbentrop Pact divided Poland, triggering the war. Stalin's purges decimated the Soviet officer corps, and his disastrous Winter War against Finland convinced Hitler that Soviet military forces could be quickly defeated, prompting Operation Barbarossa.

The Russian people's tremendous suffering and heroism in World War II occurred despite, not because of, Stalin's leadership. This inconvenient truth remains absent from Putin's narrative, which requires authoritarian leaders to appear as guarantors of national security rather than sources of catastrophic vulnerability. The omission reveals a fundamental weakness in Putin's ideological system: it cannot acknowledge how authoritarian consolidation may create the conditions for disaster rather than preventing it.

THE GLOBAL CONTEXT: TWENTY-FIRST CENTURY AUTHORITARIANISM

Putin's Russia represents one variant of twenty-first-century authoritarianism, distinct from but interactive with Chinese, Iranian, Venezuelan, and other authoritarian models. Rather than globalization leading to democratization as 1990s optimists predicted, authoritarian powers used globalization to enhance their ability to operate within and against liberal democratic orders.

Each authoritarian approach differs according to national circumstances and leadership ideology. Putin's narrative about Russia as a great Christian Orthodox power building a global tent for Russians sits awkwardly alongside support for Iran's theocracy or China's atheistic communist party. However, these authoritarian powers can work toward a broader common goal: making the world safer for authoritarianism by challenging and undermining liberal democratic norms and institutions.

Putin made clear from his first days in power that Russia would not become another Western liberal democracy. His December 1999 millennium speech established that Russia would chart its own course, with state power as the primary guarantor of order and driver of change. Over two decades, he delivered on this promise, building a centralized authoritarian system that improved many Russians' living standards while systematically eliminating political opposition and independent institutions.

Vladimir Putin's transformation of Russia from the chaos of the 1990s to an authoritarian power seeking to overthrow the Western order represents one of the twenty-first century's most significant political developments. Through a combination of ruthless consolidation of internal power, exploitation of energy wealth, and opportunistic international maneuvering, Putin positioned Russia to challenge the Western-dominated order despite relative economic and military weakness compared to the United States, China, and the European Union collectively.

His narrative of Russian victimization by the West, the need for strong state power, and Russia's destiny as a great power resonated with citizens who experienced the turbulent 1990s. By selectively leveraging Russian and Soviet history while promoting conservative values and Orthodox Christianity, Putin forged a sense of national purpose and identity absent since Soviet collapse.

However, Putinism's contradictions and limitations suggest uncertain long-term viability. The system's economic weaknesses, dependence on energy exports, demographic challenges, and reliance on personal loyalty networks rather than robust institutions create vulnerabilities. Putin's aggressive foreign policy, particularly in Ukraine, triggered Western sanctions and military buildups that constrain Russian options. The relationship with China presents both opportunities and risks as Chinese power grows along Russia's extended eastern border.

Most fundamentally, Putin's narrative cannot resolve the tension between authoritarian stability and authoritarian vulnerability to catastrophic leadership errors. By concentrating power and eliminating institutional checks, Putin created a system capable of decisive action but also prone to strategic miscalculation with limited internal mechanisms for correction.

The patriotic rhetoric and nationalist identity he has cultivated have been internalized by millions of Russians, particularly younger generations who experienced only Putin's rule. The Putin era demonstrates that the post-Cold War assumption of inevitable liberal democratic expansion was premature.

Putin's Russia represents a distinctly twenty-first-century phenomenon: a revanchist authoritarian power seeking recognition as a great power, willing to use military force to defend perceived vital interests, and capable of sophisticated diplomatic and information operations to exploit Western divisions while lacking the comprehensive power to reshape global order on its own terms.

But Putin's own ambition to restore the Russian empire through the direct invasion of a country that he does not recognize as a real country is leading to the opposite of what he has aspired to. And leaving the country as the poor cousin to Global China is something neither the Tsars nor today's Russian nationalists aspire to.

CHAPTER 2
THE RUN-UP TO PUTIN'S 2022 INVASION OF UKRAINE

In early 2022, I wrote a series of articles providing a timeline starting in 2021 where Putin and his regime have made clear their intentions with regard to Ukraine. The signs were clearly there that President Putin had turned a corner with regard to the West, and was focusing on a more direct approach towards confrontation with the West.

He had steered the country towards greater economic independence and had already learned to live with sanctions. He had made it very clear that Russia was at war with the West, hybrid and otherwise, and this level of conflict could be a surprise, but not his underlying approach.

The full series of articles can be found in chapter three of my 2025 edited book entitled, *The Biden Administration Confronts Global Change.* Here I provide a condensed version of my look at the 2021 timeline.

PACIFIC EXERCISES BY RUSSIA
AND THE LACK OF RESPONSE

While visiting Hawaii in August 2021 to consult with U.S. military commands, I learned of a major Russian military exercise that had occurred in the Pacific. This exercise signaled something far more significant than the Obama-era dismissal of Russia as merely a "regional power."

In June 2021, the Russian Ministry of Defense publicized exercises off Hawaii where they practiced destroying aircraft carrier strike groups and delivering simulated cruise missile strikes against critical military infrastructure. The exercise divided Pacific Fleet forces into opposing tactical groups, with operations conducted approximately 2,500 miles southeast of the Kuril Islands. Russian bombers spent over 14 hours airborne covering 10,000 kilometers, while Tu-142M3 aircraft were escorted by MiG-31BM interceptors. Anti-submarine warfare operations involved six Il-38 aircraft operating over the Sea of Okhotsk and Pacific Ocean.[1]

Notably, Russian forces operated within 20-30 nautical miles of Hawaii's coast. The exercise's strategic value extended beyond immediate military rehearsal. It provided intelligence directly useful to China regarding U.S. Navy capabilities in any future Taiwan contingency. The political signaling was unmistakable: Russia was demonstrating global reach and willingness to challenge American power far from traditional theaters.

The absence of any public U.S. response was puzzling. As Thomas Newdick observed, "the Pentagon's silence regarding this unprecedented threat near Hawaiian waters was baffling. This non-response reflected the atrophy of U.S. nuclear crisis management skills and the erosion of capabilities for engaging authoritarian regimes, a weakness not lost on Moscow."[2]

THE NEW RUSSIAN MILITARY DOCTRINE AND PUTIN'S HISTORICAL VISION

In July 2021, Putin's administration clarified its position through two critical documents. On July 2, Russia published its new military doctrine, ironically, the actual date the U.S. Declaration of Independence was signed. This document declared permanent conflict with the West, framing it as cultural and political warfare against Western values and institutions, particularly multinational corporations and information firms threatening Russian society.

The doctrine represented a clear break from 2015's national security strategy. It called for returning to traditional Russian values portrayed as morally superior to a decaying West. The document projected escalating confrontation with Western governments, economies, and cultural institutions. Yet when I inquired among Western decision-makers whether they had read this document, something mandatory during my government service in the early 1980s, most asked "what document?"

Putin reinforced this message with his July 2021 essay on the historical unity of Russians, Ukrainians, and Belarusians. He declared them all descendants of Ancient Rus, bound by shared language, economic ties, the Rurik dynasty, and Orthodox faith. Kiev held the dominant position as "the mother of all Russian cities." Putin explicitly rejected the concept of "Ukraine is not Russia," stating that the West's attempt to transform Ukraine into an "anti-Russia" was unacceptable.

Putin argued that the West had systematically pushed Ukraine to curtail economic cooperation with Russia while refusing dialogue in a Ukraine-Russia-EU format. He characterized Western policy as dragging Ukraine into a dangerous geopolitical game, turning it into a barrier and springboard against Russia.

These documents should have triggered alarm bells throughout the West, but they preceded actual conflict in the Black Sea that underscored Russia's seriousness.

EXERCISES, CONFLICT, AND
INFORMATION WAR IN THE BLACK SEA

From June through August 2021, the Black Sea became the theater where Russian intentions crystallized. During the Sea Breeze 2021 exercise, a multinational maritime drill co-hosted by U.S. Sixth Fleet and the Ukrainian Navy since 1997, Russia demonstrated its comprehensive approach to modern conflict.

The most significant incident involved HMS Defender's passage through international waters off Cape Fiolent near Crimea. Russia claimed it fired warning shots and dropped bombs to deter the British destroyer, though the UK disputed this account. This incident became the anchor for a sophisticated Russian information warfare campaign that followed their established pattern: gradual suspense building, aggressive media and social network attacks, and finally normalization by pushing acceptance of Russian conditions at the highest political levels.

As Ukrainian analyst Liubov Tsybulska documented, Russian operations were complex and coordinated, involving all major intelligence agencies, targeting multiple audiences simultaneously with consistent, well-tested messages across diplomatic, media, and social network channels.[3] The day before the HMS Defender incident, Putin declared NATO was breaking promises about expansion. Hours after the incident, Russia's U.S. embassy characterized the exercises as having "aggressive character." Russian intelligence then distributed "playbooks" to media, bloggers, and experts for disseminating key narratives.

Putin used his annual phone-in to reassure Russians that sinking HMS Defender wouldn't trigger world war because "the other side knows they cannot win such a war." He claimed Russia had conducted massive exercises in March-April 2021 involving over 300,000 soldiers, demonstrating capability to fight and win a conventional regional war with Ukraine while prepared for global nuclear conflict if escalation occurred.

Ukraine's Defense Ministry reported Russian hackers attacked the Ukrainian Navy website to spread disinformation about Sea Breeze, with the "entire Kremlin propaganda machine" involved.

Russian journalist Pavel Felgenhauer noted that Putin claimed Russian exercises had caused Western distress, stating he ordered gradual withdrawal from Ukraine's border, only to have the West respond with Sea Breeze 2021.[4]

These actions surrounding the exercise should have made clear that some form of action against Ukraine was imminent. Russia was conducting 21st century political-military operations characterized by escalation and de-escalation to achieve tactical and strategic advantage.

THE STRATEGIC VACUUM CREATED BY AFGHANISTAN WITHDRAWAL

The Biden Administration's August 2021 withdrawal from Afghanistan profoundly affected Putin's risk calculations. The unilateral decision to end the NATO mission created direct consequences for the global strategic situation. Following the chaotic withdrawal, Russia moved into Central Asian states, re-engaging in the power vacuum created by American departure.

For Putin, intent on shaping a new Russian empire, Central Asia's reopening accelerated imperial appetites. The withdrawal contradicted the persistent Western view that Putin led merely a decaying state.

As James Durso observed, the U.S. retreat put Central Asia on the front lines against instability. Central Asian republics, still completing state formation begun thirty years earlier with the Soviet Union's collapse, now faced a threatening Afghanistan while their primary benefactor demonstrated willingness to abandon long-term commitments.[5]

The readiness to walk away from $2.3 trillion and over 2,300 American deaths caused profound loss of confidence in U.S. assurances. Russia's "menacing embrace" would be rebranded as "standing shoulder to shoulder against instability and extremism," creating opportunities to draw all five Central Asian states into the Collective Security Treaty Organization and Eurasian Economic Union. Washington's distance from the region, previously allowing it to serve as regional balance, now worked against American interests, as Moscow had everything at stake.

The spectacle of a pandemic-weakened American military defeated by the Taliban raised fundamental questions about U.S. capability for peer warfare, particularly against nuclear-armed powers willing to use weapons as part of promoting authoritarian global order. This reality would frame Putin's assessment of Western resolve regarding Ukraine.

THE BIDEN ADMINISTRATION
EMBRACES UKRAINE

Against this backdrop of perceived weakness, the Biden Administration embraced Ukraine through partnerships that created dangerous mismatches between commitment and capability.

As historian Robert Service observed, two key blunders caused the Ukraine War. First, on November 10, the U.S. and Ukraine signed a Charter on Strategic Partnership asserting American support for Ukraine's NATO membership aspirations or "the last straw" for Putin, who immediately began preparing military operations.[6]

The situation was worse than it appeared. Ukraine would not receive NATO membership; even EU accession was unrealistic. NATO forces would not be based in Ukraine, particularly given U.S. military uncertainty after Afghanistan. The Russians represented a nuclear-armed brutal power, a different league entirely. Delusional globalization thinking that trade resolves conflicts remained strong in the West. How could an administration unprepared to confront a nuclear power defend a nation outside its alliance structure?

The timing was remarkable: Biden was unilaterally ending NATO operations in Afghanistan while embracing Ukraine, which desperately hoped the administration defeated by the Taliban would confront a nuclear power.

On September 1, 2021, the Biden Administration declared that sovereign states have rights to make decisions and choose alliances, that America's support for Ukrainian sovereignty was unwavering, and that Russia's aggression had claimed over 14,000 Ukrainian lives. They announced $60 million in security assistance and finalized a Strategic Defense Framework.[7]

The November 10 Charter was even more explicit, supporting Ukraine's NATO aspirations based on the 2008 Bucharest Summit Declaration. It pledged to support Ukraine's efforts countering armed aggression, maintaining sanctions against Russia until Ukraine's territorial integrity was restored, and refusing to recognize Crimea's annexation. The Charter endorsed enhanced defense cooperation, Black Sea security, intelligence sharing, and Ukraine's status as a NATO Enhanced Opportunities Partner.[8]

Comparing Russian July statements with these American commitments revealed a predictable collision course. The fundamental question was whether the U.S. and NATO practiced peer warfare arts against nuclear-armed powers willing to use weapons for promoting authoritarian order. A pandemic-weakened military defeated by the Taliban was rebuilding credibility.

Even more critical was the strategy for supporting states outside formal alliance structures whose independence remained crucial, a question requiring focus on creating "hedgehog states" with sufficient defensive capability to deter autocratic aggression. This required modern air defense and counter-air capabilities, not merely Javelins distributed during crises.

PUTIN'S VALDAI DISCUSSION CLUB ADDRESS

At the 18th annual Valdai Discussion Club meeting in October 2021, Putin provided his comprehensive worldview. His narrative was consistent: the Soviet Union's collapse was a global catastrophe. America attempted exploiting this collapse for hegemonic dominance but failed. The world was transitioning toward new power balance where Russia claimed its rightful place.

Putin characterized contemporary transformations as unprecedented: "This is not simply a shift in the balance of forces or scientific and technological breakthroughs, though both are taking place. Today, we are facing systemic changes in all directions." He emphasized internal Western contradictions, conflicts over sexuality, social justice, and national legacies, suggesting the United States was weakened by deep moral divisions. He compared Western "progressives" to Bolsheviks in their destructive cultural impact.

Putin rejected Western social progress prescriptions: "After the 1917 revolution, the Bolsheviks, relying on Marx and Engels, also said they would change existing ways and customs and not just political and economic ones, but the very notion of human morality and foundations of a healthy society." He cited destruction of age-old values, religion, family rejection, and informing on loved ones as historical parallels to contemporary Western trends. The Afghan war outcome further demonstrated Western weakness and declining dominance.

When asked about Ukraine and Defense Secretary Austin's visit promising NATO membership prospects, Putin responded forcefully. He recounted broken promises about NATO expansion, infrastructure approaching Russian borders, and ABM systems in Poland and Romania capable of deploying Tomahawks with simple software changes. He noted Western silence to Russia's offer not to deploy medium and short-range missiles if reciprocated. When Austin opened doors for Ukrainian NATO membership, Putin asked: "What should we do in this situation?" His question implied the answer he was preparing.

THE MIGRANT BATTERING RAM

Throughout fall 2021, migration became another weapon in Putin's arsenal. Driving people from Syria into the Middle East and Europe provided continuous disruption. Belarus's operational territory takeover in the Ukraine crisis run-up was evident through orchestrated migrant flows. During my fall 2021 Poland visit, Poles clearly understood the Belarus-Russian migrant crisis represented hybrid war, not humanitarian emergency.

At the Warsaw Defense 24 Conference in September which I attended, the "hybrid war" character of Polish border events was discussed extensively. When the Territorial Forces Commanding General discussed forming a new brigade for border security, clear sense emerged of direct threats requiring action. Poland remembered that border incursions historically preceded territorial breaches and ramped-up conflict. This aligned with permanent war against the West described in Russia's July 2021 military doctrine.[9]

Historical parallels resonated deeply. Germany's 1939 Poland attack began with staged provocations, including the Gleiwitz radio station incident where Germans posing as Poles broadcast anti-German messages, providing Hitler justification for invasion.

As museum director Andrzej Jarczewski noted: "The provocations in Gliwice and some other places were necessary to allow Hitler to make his speech, to say 'we are innocent, the Poles started this war.'" The Russian-Belarusian provocations reminded Poles of history they refused to repeat.[10]

CONCLUSION: SIGNALS IGNORED

The 2021 timeline reveals an unmistakable trajectory toward military confrontation marked by deliberate Russian signaling, Western ambiguity, and fundamental failure recognizing Putin's strategic vision coherence.

The February 2022 invasion was not a sudden policy pivot but rather the logical culmination of a carefully orchestrated campaign combining doctrinal statements, military exercises, information warfare, and geopolitical maneuvering.

From June Pacific exercises demonstrating global reach through July's doctrinal publications establishing permanent conflict framework, from Black Sea confrontations showcasing hybrid warfare capabilities through Afghanistan withdrawal signaling American weakness, from contradictory Ukrainian commitments creating capability-commitment mismatches through Valdai addresses articulating civilizational narratives, and finally through migration weaponization exposing European vulnerabilities, Putin consistently telegraphed intentions through doctrine, rhetoric, and action.

The problem was not signal absence but Western inability integrating signals into coherent threat assessments. Each event appeared manageable in isolation, but together formed unmistakable escalation patterns.

Western failure responding effectively reflected deeper pathologies: atrophied nuclear crisis management skills, wishful thinking about economic interdependence stabilizing effects, internal political divisions, and fundamental uncertainty about supporting states outside formal alliance structures against nuclear-armed adversaries.

The fundamental question emerging from this timeline is not why Russia invaded Ukraine, but why Western policy communities failed anticipating and deterring an invasion so clearly signposted.

The answer lies partly in disconnects between authoritarian strategic thinking reality and liberal international order theory assumptions.

It also reflects costs of treating diverse Russian behaviors, military exercises, doctrinal statements, information operations, energy coercion, migration manipulation, as separate phenomena rather than integrated comprehensive strategy elements.

Defending sovereignty and deterring aggression requires not just military capabilities but intellectual frameworks understanding how authoritarian powers prosecute political warfare across multiple domains simultaneously.

The failure preventing Ukraine's 2022 crisis began not with any single decision but with systematic failure reading comprehensive signals Putin sent throughout the preceding year.

Each event was a data point in a larger pattern that, properly interpreted, revealed intentions Western policy makers proved unable or unwilling to acknowledge until military columns crossed Ukrainian borders.

SECTION II
THE RUSSIAN DYNAMIC

CHAPTER 1
THE PATH NOT TAKEN

In the rapidly shifting landscape of post-Cold War Europe, few diplomatic initiatives were as ambitious or ultimately as successful as NATO's Partnership for Peace (PfP) program. Launched in 1994, the PfP represented a significant approach to reworking European security, creating a framework for cooperation between NATO and non-member states without the full commitments and guarantees of Alliance membership.

Yet today, as the war in Ukraine grinds on and tensions between Russia and the West have reached their highest levels since the Cold War, the history of this remarkable program has been largely forgotten, particularly by those who once participated in it most actively.

This historical amnesia obscures something crucial: the current conflict was not inevitable. Russia once chose cooperation, participated actively in Western security frameworks, and had genuine opportunities to pursue cooperation over domination.

Understanding why those opportunities were rejected reveals that the war in Ukraine reflects not just Western policies, but fundamental choices made by Putin about how Russia would relate to its neighbors and the broader European community.

WHEN COOPERATION WAS POSSIBLE

The Partnership for Peace emerged from the complex geopolitical realities of the early 1990s. President Bill Clinton unveiled the program at the NATO summit in Brussels on January 10, 1994, offering a path for engagement that wouldn't trigger Russian fears of encirclement while still providing security cooperation for vulnerable post-Communist states. The framework was built around voluntary participation, self-differentiated cooperation, and practical military activities focused on peacekeeping, humanitarian operations, and democratic military reform.

Russia formally joined PfP on June 22, 1994, and for over a decade engaged actively and constructively in the program's activities. Russian forces trained alongside NATO troops in numerous exercises, including the major Partnership for Peace exercise in Poland in 1996. Russian officers attended NATO staff colleges and participated in planning processes. Most significantly, Russian forces served under NATO command in Bosnia, a development that would have been unthinkable just years earlier.

The cooperation was extensive and substantive. Russian and NATO forces worked together on nuclear security, defense conversion, military-to-military contacts, and democratic military reform. The Russian sector in Bosnia, centered around Ugljevik, became a symbol of post-Cold War cooperation possibilities. Russian peacekeepers worked effectively within the broader NATO framework while maintaining their national character, helping legitimize the international intervention among Orthodox Serbs.

The practical benefits were substantial. By the early 2000s, NATO and Russian forces had developed unprecedented interoperability. Joint training exercises had created personal relationships between officers that transcended national boundaries. Intelligence sharing on terrorism and proliferation had become routine. Most importantly, a generation of Russian military officers had been exposed to NATO's military culture and democratic values.

This entire history has been systematically erased from contemporary Russian political discourse. Vladimir Putin's narrative of NATO as an inherently hostile organization bent on Russia's encirclement makes no mention of the decade-plus period when Russian forces trained alongside NATO troops and Moscow actively participated in shaping European security architecture.

This selective amnesia obscures what may have been the most successful experiment in East-West military cooperation since 1945 and the fundamental choice that ended it.

THE ENERGY WEALTH WINDOW

Beyond military cooperation, Russia in the early 2000s found itself in a historically unique position. High oil and gas prices generated unprecedented wealth, while Europe's growing energy needs created obvious opportunities for deeper integration.

Countries like Norway demonstrate how energy-rich nations can maintain full sovereignty while building intimate economic and institutional ties with European partners. Russia possessed far greater resources and geographic advantages than Norway, along with a highly educated population and significant technological capabilities inherited from the Soviet era.

Yet Vladimir Putin chose a fundamentally different path. Rather than viewing energy exports as a foundation for mutual prosperity and integration, his administration increasingly treated them as instruments of geopolitical leverage.

The pattern emerged early: energy cutoffs to Ukraine in 2006 and 2009, supply disruptions affecting multiple European countries, and the explicit linkage of energy contracts to political compliance. This approach revealed a conception of international relations based on dominance rather than partnership.

The contrast with alternative approaches is stark. Germany's postwar integration with Western Europe, despite initial resistance and suspicion, ultimately created unprecedented prosperity and security. Japan's relationship with the United States, built on shared institutions and economic interdependence, transformed a former enemy into a cornerstone ally.

Russia possessed the resources and capabilities to pursue similar integration with Europe, potentially creating a continental economic zone that could have rivaled any other global bloc.

THE IMPERIAL FRAMEWORK

Understanding why Russia rejected these opportunities requires examining the ideological framework that guided Putin's thinking. From his early years in power, Putin articulated a vision of Russia as a distinct civilization with natural rights to influence over neighboring territories. This wasn't merely about security concerns or fears of encirclement. It reflected a fundamental belief that certain regions belonged within Russia's sphere of influence regardless of their populations' preferences.

This imperial logic manifested in consistent patterns across multiple countries. In Georgia, Russia supported separatist regions and eventually recognized their independence after the 2008 war, the same year that marked a decisive break in Russia-NATO cooperation. In Moldova, it maintained military forces in Transnistria despite lacking legal basis. In Belarus, it intervened decisively to prevent Alexander Lukashenko's ouster after fraudulent elections. In Kazakhstan, it deployed troops to help suppress popular protests in 2022.

These interventions shared common features: they prioritized regime stability over democratic legitimacy, used military force to prevent political change, and treated sovereignty as conditional on alignment with Russian preferences. This approach was bound to generate resistance from populations seeking different political arrangements, regardless of NATO's actions or Western policies.

The 2008 Bucharest NATO summit, where Georgia and Ukraine were promised eventual membership, is often cited as a turning point. But the underlying shift had begun earlier.

Putin's opposition to democratic movements in neighboring countries intensified even before serious NATO discussions. The

2004 Orange Revolution in Ukraine triggered alarm in Moscow not primarily because it brought Ukraine closer to NATO membership which remained deeply unpopular among Ukrainians at the time but because it demonstrated that Ukrainians could organize effective resistance to Russian-preferred outcomes and aspire to European integration.

THE REAL THREAT: NOT NATO, BUT THE EU

Here lies the critical insight that much Western analysis misses: It is the European Union, not NATO, that represents the existential threat to Putin's vision of restored Russian empire. A successful, democratic, prosperous Ukraine integrated with European institutions would demonstrate that former Soviet republics could chart independent courses while maintaining positive relationships with Russia. It would prove that the "civilizational" arguments for Russian hegemony were false, and that alternative models of development were not only possible but superior.

The European Union's appeal to post-Soviet countries wasn't primarily about ideology or security guarantees. It was about access to markets, investment, and economic relationships based on predictable rules rather than political whims.

Russia's economic approach to its neighbors consistently prioritized control over prosperity. Trade wars with Poland, the Baltic states, Georgia, and Moldova followed similar logic: economic tools were subordinated to political objectives, creating resentment and driving these countries toward Brussels. Rather than using trade and investment to create mutual benefits, Russia repeatedly weaponized economic relationships, gas price manipulation, trade restrictions, discriminatory policies designed to punish political choices rather than address commercial concerns.

THE DEMOCRATIC CHALLENGE

Perhaps most fundamentally, Putin's system could not tolerate successful democratization in neighboring countries because it exposed the weaknesses of his own model. The color revolutions in Georgia (2003), Ukraine (2004), and Kyrgyzstan (2005) demonstrated that post-Soviet populations could organize effective challenges to authoritarian rule. These movements succeeded not because of Western manipulation though they certainly received external support but because they addressed genuine grievances about corruption, economic stagnation, and political repression.

Russia's response was revealing. Rather than addressing similar problems within Russia or offering alternative governance models, Putin's administration consistently supported efforts to reverse democratic gains. This included backing Viktor Yanukovych in Ukraine despite his obvious weaknesses, supporting authoritarian consolidation in Belarus and Kazakhstan, and intervening directly when peaceful transitions seemed likely.

This pattern suggests the conflict over Ukraine was never really about NATO bases or military threats. It was about competing models of political organization and development. Putin's system required neighboring countries to remain weak, divided, and dependent to validate its own approach and prevent demonstration effects that might inspire Russian citizens.

WHY ACCOMMODATION FAILED

Understanding these deeper issues helps explain why the conflict escalated despite repeated attempts at accommodation. Western policymakers frequently assumed that addressing Russia's stated security concerns would reduce tensions.

But Putin's actual concerns were not primarily about military deployments or alliance structures. They were about the political and economic success of neighboring democracies.

This explains the futility of proposals for Ukrainian neutrality or limitations on NATO expansion. From Putin's perspective, a neutral but successful Ukraine integrated with European economic institutions would be almost as threatening as a NATO member. It would still demonstrate that former Soviet republics could prosper under different political arrangements, potentially inspiring similar transitions elsewhere.

The invasion itself revealed the bankruptcy of the imperial approach. Putin apparently believed Ukrainians would welcome Russian forces as liberators from a supposedly illegitimate government. This catastrophic miscalculation reflected years of consuming his own propaganda about Ukrainian identity and political preferences. The fierce resistance that followed demonstrated how completely he had misunderstood the society he claimed to be protecting.

ALTERNATIVE HISTORIES

Considering what might have been illuminates the extent of these missed opportunities. A Russia that had chosen cooperation over domination could have become the dominant partner in a continental European system. Russian energy resources, European technology and capital, and Ukrainian agricultural potential could have created unprecedented prosperity across the region.

The Partnership for Peace demonstrated this was possible. For a brief period in the 1990s and early 2000s, it seemed to offer a model for transcending traditional alliance structures in favor of more flexible, inclusive approaches to collective security. The success stories were real: Poland, Hungary, and the Czech Republic used PfP as a pathway to successful NATO integration, transforming their militaries and defense establishments through practical cooperation rather than confrontation.

Such an arrangement would have required Russia to accept certain constraints: respect for democratic processes, adherence to international law, and genuine rather than performative anti-corruption efforts. But these constraints would have been far less burdensome than the costs of the current conflict: international isolation, economic sanctions, military losses, and the transformation of Ukraine from a potential partner into an implacable enemy.

The tragedy is that this alternative was readily available. The European Union repeatedly signaled its willingness to develop closer relationships with Russia based on shared values and mutual benefit. Individual European countries, particularly Germany, went to extraordinary lengths to maintain positive relationships even after the annexation of Crimea. Russia's rejection of these opportunities in favor of confrontation reflects choices, not inevitabilities.

BEYOND BALANCE OF POWER

Contemporary discussions of the Ukraine conflict often default to balance-of-power analysis, treating it as a predictable result of major power competition. This framework, while offering some insights, obscures the role of ideology and domestic political systems in shaping international behavior.

Putin's Russia was not simply responding to external threats or pursuing traditional geopolitical objectives. It was defending a particular model of governance and development that required neighboring countries to remain weak and dependent. This model was fundamentally incompatible with the European approach of voluntary integration based on shared institutions and democratic governance.

The conflict thus represents not just a clash between Russia and the West, but between different concepts of how societies should be organized and how international relationships should function. Putin's approach assumes that strong states naturally dominate weak ones, that sovereignty is conditional on power and that international law serves primarily to legitimize the preferences of dominant actors. The European alternative, however imperfectly realized, envisions relationships based on mutual benefit, respect for democratic choices, and gradual convergence around shared standards.

LESSONS FROM
FORGOTTEN COOPERATION

The current amnesia about Partnership for Peace serves no constructive purpose. Acknowledging what cooperation achieved does not mean ignoring why it failed. The program succeeded because it offered tangible benefits to all participants, avoided zero-sum thinking, and built relationships gradually through practical cooperation rather than grand political gestures.

Personal relationships and professional trust built through joint exercises proved more durable than many formal agreements. Russian and NATO officers who trained together in the 1990s maintained professional respect even as their governments moved toward confrontation. These relationships demonstrated that practical cooperation can create political possibilities that formal diplomacy cannot achieve alone.

The program also showed that military cooperation can promote democratic values and governance reforms. The requirement for civilian control of the military, transparent defense budgeting, and adherence to international humanitarian law became standard expectations. Many countries reformed their defense establishments primarily to meet PfP standards.

Yet PfP also revealed limitations. While the program created unprecedented cooperation between former adversaries, it could not ultimately overcome fundamental geopolitical disagreements about European security architecture and spheres of influence. The question remains: were these disagreements truly irreconcilable, or did specific choices by Russian leadership make them so?

Recognizing Russia's agency and choices doesn't excuse Western mistakes or absolve NATO of responsibility for its own decisions. The expansion of the alliance, particularly promises made to Ukraine and Georgia in 2008, clearly contributed to escalating tensions. Western support for color revolutions, while based on genuine democratic values, undoubtedly alarmed Russian leaders.

But focusing exclusively on Western actions risks treating Russia as a passive respondent rather than an active participant with its own objectives and strategies. Putin's Russia made deliberate choices to prioritize control over prosperity, confrontation over cooperation, and imperial nostalgia over democratic development. These choices shaped the trajectory toward conflict as much as any Western decisions.

Understanding this dynamic is crucial for developing effective responses and preventing future conflicts. Accommodation strategies that ignore Russia's fundamental ideological commitments are likely to fail, while confrontational approaches that treat military deterrence as sufficient may miss opportunities for engagement where interests actually align.

The war in Ukraine represents the culmination of decades of choices by both Russian and Western leaders. While Western policies certainly contributed to escalating tensions, Russia's rejection of alternative approaches to regional relationships made conflict increasingly likely regardless of NATO's specific actions.

Moving forward requires acknowledging that this conflict reflects deeper disagreements about political organization, economic development, and international law. These disagreements cannot be resolved through territorial adjustments or security guarantees alone. They require sustained engagement with underlying questions about how societies should govern themselves and relate to their neighbors.

The ultimate resolution will depend not just on military outcomes or diplomatic negotiations, but on whether Russia can develop alternative approaches to regional relationships that respect sovereignty while addressing legitimate security concerns. Such evolution would require fundamental changes in Russian political culture and strategic thinking, changes that may only be possible after the current system exhausts itself through its own contradictions.

Until then, the West faces the challenge of defending democratic values and international law while remaining open to genuine Russian interest in alternative arrangements. This balance requires both strength and wisdom, recognizing that lasting peace depends not just on containing Russian power, but on creating conditions where Russia can pursue prosperity and security through cooperation rather than domination.

The men and women who worked to make Partnership for Peace successful demonstrated that alternatives to permanent confrontation are possible. Their work deserves to be remembered, not as naive idealism, but as proof that even the most entrenched conflicts can be transformed through patient, practical cooperation.

As Europe grapples with renewed division, the PfP experience offers both hope because it shows cooperation between former enemies is possible and warning, because it demonstrates how quickly such cooperation can unravel when underlying political disagreements are left unresolved.

Only by remembering what cooperation achieved and honestly assessing why it failed can we hope to build the security architecture that Europe needs for the twenty-first century.[1]

I would add that this discussion is not an abstract one for me as I worked within the Partnership for Peace program, particularly with regard to Bosnia. I saw first hand and participated in the support of the program where real differences were confronted and dealt with. It was not about wishing away differences but confronting them.

But for Putin such cooperation exposed the nature of his regime, the level of cooperation of his system of government and put in bold relief his attachment to empire over engagement in the 21st century world.

CHAPTER 2
STRATEGIC MISCALCULATIONS

When Vladimir Putin launched his "special military operation" against Ukraine on February 24, 2022, he envisioned a swift victory that would restore Russia's great power status, prevent NATO expansion, and reassert Moscow's dominance over its sphere of influence.

Instead, the invasion has triggered a cascade of consequences that have systematically weakened Russia across multiple dimensions: militarily, scientifically, economically, and geopolitically. What emerges from examining this conflict is a stark illustration of how authoritarian aggression can backfire spectacularly, creating precisely the threats it purports to address while dismantling the very foundations of national strength.

Putin's decision to launch a full-scale invasion of Ukraine in February 2022 represented the culmination of his growing frustration with what he viewed as Western encirclement. Yet this decision, rather than resolving Russia's security dilemma, has created exactly the militarized and hostile NATO that Putin claimed to fear.

Within days of the Russian invasion, NATO underwent a transformation that previous decades of gradual expansion had failed to achieve. The alliance displayed a unity and sense of common purpose absent since the height of the Cold War. Germany's response was perhaps the most dramatic. Chancellor Olaf Scholz announced a "Zeitenwende" (turning point) that would see German defense spending increase to over 2 percent of GDP, with a special €100 billion fund for military modernization. The nation that had effectively disarmed itself committed to building one of Europe's largest and most capable militaries.

Similar transformations occurred across the alliance. Poland announced plans to double the size of its military and become home to the largest land force in Europe outside of Russia. The United Kingdom reversed years of defense cuts with significant increases in military spending. Even traditionally neutral nations like Switzerland began reconsidering their defense policies.

Perhaps no development better illustrates the self-defeating nature of Putin's strategy than the decision of Finland and Sweden to abandon their long-standing neutrality and join NATO. These nations had maintained their non-aligned status throughout the Cold War, even as the Soviet Union posed a far more credible military threat than the Russia of 2022. Finland's decision was particularly significant given its 830-mile border with Russia. Swedish neutrality had been a cornerstone of European security architecture since the early 19th century, surviving both World Wars and the Cold War. Yet Putin's aggression accomplished what centuries of major power competition had not.

The alliance response has gone far beyond increased defense spending and new membership. NATO has fundamentally restructured its force posture, establishing new forward-deployed battle groups in Eastern Europe and moving from symbolic "tripwire" forces to brigade-sized formations capable of meaningful territorial defense. Air defense systems, largely neglected during the post-Cold War focus on counterinsurgency operations, have become a priority across the alliance.

Putin's approach reveals a classic self-created security dilemma, a situation in which actions taken to enhance one's own security inad-vertently threaten others, leading them to take countermeasures that ultimately reduce the security of all parties.

At every stage, Putin's actions have created precisely the security challenges he claimed to be addressing. His 2014 seizure of Crimea ended the post-Cold War European security consensus and began NATO's reorientation toward territorial defense. His 2022 invasion completed this transformation, creating a militarized and united alliance that poses a far greater threat to Russian security than the scattered and underfunded NATO of previous decades.

THE SCIENTIFIC EXODUS: DISMANTLING RUSSIA'S INTELLECTUAL FOUNDATION

Beyond the immediate military and geopolitical consequences, Putin's invasion has triggered one of the most devastating scientific brain drains in modern history, systematically dismantling the math-ematics and physics capabilities that Russia had built over centuries.

Russia has long prided itself on its scientific heritage. From Dmitri Mendeleev's periodic table to Andrei Kolmogorov's contribu-tions to probability theory, from Lev Landau's groundbreaking work in theoretical physics to the Soviet space program's early triumphs, Russian science has produced world-changing discoveries. The coun-try's mathematical traditions, in particular, have been legendary, with Soviet mathematicians dominating international competitions and making foundational contributions to fields ranging from topology to number theory.

Today, that legacy lies scattered across European and American universities and research institutes where Russian scientists have sought refuge from Putin's increasingly authoritarian regime. According to reports from multiple sources, at least 2,500 scientists have fled Russia since February 24, 2022, with experts describing the phenomenon as a "disaster" for Russian science. This figure likely represents a conservative estimate, as many departures go unreported.[1]

Among those who have left are at least 34 physicists and mathematicians from Russia's most prestigious institutions. These are not marginal figures but acclaimed scientists with established careers. Up to a quarter of the departing scientists have citation H-Index scores of 10 or higher, a metric indicating successful careers spanning decades of research experience. Many maintained ties to foreign universities, making their departure not just a loss of individual talent but a severing of crucial international networks

The brain drain has struck hardest at the most prestigious and internationally connected universities. The Higher School of Economics (HSE), which Vladimir Putin himself had praised in 2010 as "cutting-edge in every respect," has lost approximately 700 faculty members since the war began, causing it to plummet almost 100 spots in global rankings, from 305th to 399th position.[2]

The HSE's decline began even before the invasion, as authorities increased pressure on faculty members who showed insufficient loyalty to the regime. The university received a new rector in 2021 who launched a gradual purge of faculty members. Ilya Inishev, a Doctor of Philosophy who worked at HSE from 2010 until 2022, was eventually dismissed for "serious damage" his antiwar comments "inflicted on the university's reputation" and moved to Germany in April 2023.

Research by Novaya-Europe identified at least 270 university staff members from Moscow and St. Petersburg's high-ranking universities who have severed ties with Russia since the war broke out. Among these, 195 are considered Russian scientists, while the rest are foreigners who had been working in Russia. These figures represent only the verified cases from open sources. The actual number is likely much higher.[3]

Putin's invasion also provoked an unprecedented international scientific boycott that has systematically isolated Russian institutions from global research networks. One of the most symbolic casualties was the $300-million Skoltech program, a joint initiative between MIT and Russian partners that represented one of the most ambitious East-West scientific collaborations of the post-Cold War era.

The program was dissolved within one day of the invasion, with no foreseeable restart in the future.[4]

The European Organization for Nuclear Research (CERN), which had served as a bridge between East and West even during the darkest days of the Cold War, made the unprecedented decision to bar all Russian observers and terminate the contracts of approximately 1,000 Russian scientists, about 8% of its workforce. This move was particularly significant given CERN's historic role as a meeting place for scientists from opposing blocs, a function it had maintained even during the Soviet invasions of Czechoslovakia in 1968 and Afghanistan in 1979.[5]

What makes this brain drain particularly devastating for Russia is not just its scale but its quality. The scientists who are leaving represent the most internationally connected and productive segments of the Russian research community. Analysis of GitHub data shows that by November 2022, 11.1% of Russian developers had listed a new country, compared with only 2.8% of developers from comparable countries not directly involved in the conflict. More tellingly, the 11% of developers who left Russia had been responsible for 20% of the country's international collaborations in the software development community.[6]

This pattern reflects a broader truth about brain drain: it is not random. Those who leave tend to be better connected both domestically and internationally. In the global collaboration network, 43.0% of departing developers had ties with colleagues in other countries, compared with only 24.3% of those who remained. The same dynamics likely apply to academic researchers, meaning that Russia is losing not just individual scientists but the crucial nodes that connected Russian science to the global research community.[7]

The current Russian scientific exodus bears uncomfortable similarities to Nazi Germany's purge of Jewish scientists in the 1930s. When Hitler declared he would rid German universities of Jews even if it meant "the annihilation of contemporary German science," he achieved exactly that outcome. The 15% of German physicists who lost their jobs were the country's most productive researchers, accounting for 64% of all physics citations in Germany.

Many of these displaced scientists found refuge in Britain and the United States, where they continued their groundbreaking work. Several would play crucial roles in the Manhattan Project, a bitter irony that saw Germany's scientific talent contributing to the Allied victory over Nazi Germany. Economic research on such episodes confirms their long-lasting impact. Studies of academic emigration from Nazi Germany show that these shocks diminished local research productivity for decades. Human capital, unlike physical capital, cannot be quickly rebuilt or replaced.[8]

What makes the current Russian brain drain particularly tragic is that it appears to be accelerating due to the government's own actions. Since 2015, at least a dozen Russian physicists have been arrested on charges of "high treason" for the simple act of working with foreign colleagues or publishing in foreign journals. The Russian authorities have explicitly embraced what they call a policy of isolationism from the international scientific community, deliberately echoing the practices of the Soviet Iron Curtain era.[9]

The brain drain occurs against the backdrop of a broader economic crisis that had been undermining Russian science even before the war. Since 2008, funding for scientific research as a percentage of GDP has stagnated at around 1%. Salaries for researchers remain dismally low. A senior researcher earns an average of 26,000 rubles (about €280) per month, while a professor earns 36,000 rubles (€390).[10]

The full impact of this scientific exodus will likely take decades to manifest, but the early indicators are ominous for Russia. International collaborations with Russian scientists fell by 34% by 2024 compared to 2021 levels. More fundamentally, Russia is losing its capacity to train the next generation of world-class researchers. Graduate programs are being disrupted by faculty departures, while the most promising students are increasingly likely to seek opportunities abroad.

THE CRUMBLING EMPIRE: GEOPOLITICAL LOSSES ACROSS MULTIPLE FRONTS

Putin's invasion has not only strengthened NATO and gutted Russian science; it has also precipitated a series of geopolitical crises that are fundamentally reshaping Russia's relationships with former allies and dependencies across multiple regions.

The Baltic Success Story

The most successful example of strategic decoupling from Russia comes from the Baltic states, Estonia, Latvia, and Lithuania. Lithuania led the transformation by building the "Independence" LNG terminal in Klaipėda in 2014, despite active Russian pressure. This strategic investment enabled Lithuania to become the first European country to completely ban all Russian gas imports in April 2022.

In February 2025, the Baltic states officially severed their electricity connections with Russia and Belarus, ending more than three decades of dependence on the Soviet-era BRELL electricity network. Despite initial concerns about economic impact, the Baltic states demonstrated remarkable resilience. The overall GDP impact from severing Russian ties was estimated at less than 0.5% in each country. Lithuania emerged as the regional economic leader in 2024 with robust 2.9% growth.[11]

Kazakhstan's Strategic Pivot

Kazakhstan's shift away from Russia represents perhaps the most strategically damaging blow to Moscow's war effort. Long considered one of Russia's most reliable partners in Central Asia, Kazakhstan has systematically cut economic and military ties while pivoting toward Western partnerships.[12]

The most critical disruption involves cotton pulp, an obscure but vital commodity for Russian ammunition production. For decades, Kazakhstan and Uzbekistan supplied over 98% of Russia's imported cotton pulp, which is processed into nitrocellulose, the key component in virtually every Russian missile, drone, and artillery shell. Russia's imports of nitrocellulose and cotton pulp have been sufficient to produce the equivalent of 700,000 artillery shells annually.[13]

However, recent indicators suggest Kazakhstan is deliberately cutting back these shipments. The shift became apparent in summer 2023 when Khlopkoprom-Cellulosa, Kazakhstan's main cotton pulp producer, abruptly halted operations.[14]

Beyond cotton, Kazakhstan has systematically reduced its energy dependence on Russia, expanding use of the Baku-Tbilisi-Ceyhan pipeline to export oil via Turkey to European markets, a direct bypass of Russian territory that undermines Moscow's regional leverage. The country recently dismissed pro-Russian Defense Minister Ruslan Zhaksylykov and is now participating in NATO-standard ammunition production projects.[15]

Azerbaijan's Hardline Turn

The trajectory of Azerbaijan-Russia relations fundamentally shifted following the December 25, 2024 downing of Azerbaijan Airlines Flight 8243 by a Russian surface-to-air missile system. The incident, which killed 38 people and wounded 29, catalyzed a dramatic deterioration in relations between two countries that had maintained pragmatic cooperation for over three decades.[16]

President Ilham Aliyev's response was swift and uncompromising, demanding that Russia apologize, admit guilt, and pay compensation. The crisis reached a new nadir in June 2025 when Russian police conducted violent raids against ethnic Azerbaijanis in Yekaterinburg, resulting in the deaths of two brothers who had been beaten to death in custody. Azerbaijan's response was unprecedented: it raided the Baku office of Russian state media outlet Sputnik, arrested Russian nationals, and launched a criminal investigation into what it termed the "torture and murder with extreme cruelty" of Azerbaijani citizens.[17]

Finally, in October 2025. President Vladimir Putin has shifted his official stance regarding the downing of the Azerbaijani passenger jet, moving beyond a general apology and now publicly acknowledging Russia's responsibility for the incident. Putin has assured Azerbaijan that Moscow will provide compensation for its part in the tragedy and has committed to an objective assessment of all actions taken by Russian officials involved. This change follows months of frustration in Baku, where earlier statements from Russia were seen as wholly insufficient and evasive.[18]

Azerbaijan's boldness in challenging Russia stems partly from its strengthened position following its decisive victory over Armenia in 2023, when it recaptured Nagorno-Karabakh while Russia, preoccupied with Ukraine, failed to intervene.[19] Turkey's role has been pivotal, providing Azerbaijan with advanced military hardware and an alternative security guarantee that reduces dependence on Russian equipment and protection.[20]

Germany's Economic Devastation

Germany's relationship with Russia, built over decades through Ostpolitik and anchored in the principle of "Wandel durch Handel" (change through trade), collapsed virtually overnight following the invasion. German exports to Russia plummeted by 67% between 2019 and 2023, falling from €32 billion to just €10.6 billion. The sectoral impact was even more dramatic: exports of cars and car parts to Russia declined by 94%, machinery exports fell by 76%, and electronic products dropped by 88%.[21]

Before the war, over 50% of Germany's gas and coal, and more than 30% of its oil came from Russia. The German government moved with unprecedented speed to sever these energy ties, completely halting pipeline natural gas imports from Russia by

September 2022. Natural gas prices surged tenfold at their peak, creating cascading effects throughout the German economy. Higher energy prices and trade disruptions cost the German economy at least €100 billion, equivalent to 2.5-4% of GDP.[22]

The crisis fundamentally discredited Germany's approach to Russia that had defined its foreign policy for half a century. German distrust of Russia reached a record 90% following the invasion, and by 2024, 95% of Germans disapproved of President Vladimir Putin.[23]

China's Expanding Influence

China has become the largest trading partner and foreign investor in the Russian Far East, pouring significant funds into infrastructure, energy, agriculture, and port development. Over 53 Chinese companies operate in 18 advanced development territories and 22 ports across the Russian Far East, with planned investment exceeding 816 billion rubles.[24]

The territorial relationship between China and Russia remains troubled. Despite a 2008 agreement that divided Bolshoy Ussuriysky Island between the two countries, China's Ministry of Natural Resources published a "standard map" in August 2023 that appeared to claim the entire island as Chinese territory.[25] This economic relationship has become even more critical for Russia following the 2022 invasion and subsequent Western sanctions, with Chinese companies now accessing sectors previously closed to foreigners due to security barriers, including defense industries.[26]

The crisis reflects broader trends undermining Russian influence across the former Soviet space. Armenia has distanced itself from Russia following the Nagorno-Karabakh defeat and is now pursuing closer ties with the EU and NATO. Georgia continues its Western orientation despite Russian pressure. What was once considered unthinkable defiance of Moscow has become increasingly common as regional states recognize Russia's reduced capacity to enforce its will.[27]

THE WAR ECONOMY: A TOOL
OF DOMESTIC CONTROL

Beyond its external consequences, Putin's invasion has fundamentally transformed Russia into a war economy but not merely as a means to achieve territorial conquest. The prolonged conflict has evolved into something far more strategically valuable for Putin: a mechanism for consolidating domestic power and tightening his grip on Russian society.[28]

The war's failure to achieve its initial objectives has created an unexpected political dividend for Putin. By channeling resources away from oligarchs and forcing the population to accept a lower standard of living to fund the war machine, Putin has managed to centralize economic control to an unprecedented degree. The security services have been strengthened and expanded, ostensibly to manage the war economy but effectively to control every aspect of Russian life.

This transformation reveals a crucial paradox: military setbacks on the battlefield have translated into political victories at home. The war economy justifies increased state surveillance, provides a nationalist rallying point that suppresses dissent, and allows Putin to redistribute resources from potential rivals to his security apparatus. What began as an external campaign has become an internal consolidation project.

The war economy has created its own logic of perpetuation. Ending the conflict would not simply return Russia to its pre-2022 status quo; it would potentially unravel the entire power structure that Putin has constructed around the war effort. The centralized control over resources, the expanded security services, and the heightened state of national mobilization all depend on the continuation of the conflict.

This creates a fundamental disincentive for peace that goes beyond traditional military or territorial calculations. For Putin, the war has become less about conquering Ukraine and more about maintaining his position within Russia. The conflict provides ongoing justification for authoritarian measures that might otherwise provoke resistance from the population or elite circles.

This raises the most pressing question facing international diplomacy: is there an off-ramp that would actually interest Putin?

Any viable peace agreement would need to somehow preserve or even enhance his domestic position rather than simply address territorial disputes or security guarantees. Traditional diplomatic approaches that focus solely on military and territorial concessions may be fundamentally insufficient because they ignore the domestic political utility that the war provides.[29]

THE CRIPPLING PRICE OF IMPERIAL AMBITION

Putin's invasion of Ukraine has inflicted many costs on Russia, from economic sanctions to international isolation to military casualties. But perhaps none will prove as enduringly damaging as the comprehensive weakening of Russian power across multiple dimensions that began on February 24, 2022.

Rather than expanding the Russian empire, Putin increasingly resembles Nicholas II presiding over the systematic dismantling of Russian influence across multiple fronts. The invasion, intended to restore Russian greatness and prevent NATO expansion, has instead:

- Created the most militarized, unified, and hostile NATO since the Cold War.
- Driven Finland and Sweden into the alliance, adding 830 miles of NATO border with Russia.
- Triggered an unprecedented scientific brain drain that has gutted Russia's research capabilities.
- Severed critical economic relationships, particularly Germany's complete energy pivot away from Russia.
- Fractured Russia's influence across former Soviet states, from the Baltic to Central Asia.

- Transformed Ukraine into Europe's most heavily armed state with modern Western weapons.
- Russia economically subordinating itself to China and ceding much of the Russian Far East to de facto Chinese control.

The irony is particularly acute given Putin's stated goals. His shadow boxing with a largely benign and distracted NATO created precisely the militarized alliance he claimed to fear.[30] His suppression of scientific freedom drove away the researchers who represented Russia's future competitive advantage. His attempts to maintain imperial influence accelerated the independence of former dependencies.

What emerges from this analysis is a picture of unintended consequences on a massive scale. Putin's miscalculations, believing Ukraine would not resist, assuming NATO would remain divided, ignoring the value of scientific openness, underestimating the resolve of neighboring states. have combined to produce outcomes diametrically opposed to his intentions.

The transformation of Russia since February 2022 reveals how authoritarian decision-making, divorced from accurate intelligence and dissenting viewpoints, can lead to catastrophic strategic failures. Putin's increasingly isolated leadership style created an echo chamber where optimistic assumptions went unchallenged and contradictory evidence was dismissed or suppressed.

The full implications of these geopolitical realignments are still unfolding, but the trajectory is clear: Putin's invasion of Ukraine has not strengthened Russia's position but has instead precipitated the most significant contraction of Russian influence since the collapse of the Soviet Union itself. The regional balance of power is fundamentally shifting, with countries that once operated within Moscow's sphere of influence now demonstrating that alternatives to Russian partnership are not only possible but potentially more beneficial.

Perhaps most significantly, the war economy that Putin has

constructed around the conflict creates its own logic of perpetuation. The centralized control, expanded security apparatus, and nationalist mobilization all depend on continued conflict, making peace potentially destabilizing to Putin's domestic position. This suggests that ending the war may require addressing not just territorial disputes but the fundamental political dynamics that have made the conflict valuable to Putin's continued rule.

The tragedy extends beyond Russia's borders. Science is fundamentally a cooperative enterprise that benefits from the free exchange of ideas regardless of nationality. The barriers that now separate Russian scientists from international colleagues represent a loss for human knowledge as well as a setback for Russia specifically. The militarization of Europe, while perhaps necessary for defense, represents a diversion of resources from productive economic activity to military preparation.

Yet the responsibility for this catastrophe lies squarely with Putin's choices. By launching an unprovoked war of aggression, by persecuting scientists who maintain international contacts, by embracing isolation over cooperation, and by treating neighboring states as subordinate rather than sovereign, Russia's leadership has chosen policies that inevitably drive away talent, alienate allies, and strengthen adversaries.

The mathematicians and physicists now working in European and American universities, the former dependencies now charting independent courses, the militarized NATO alliance, all represent living proof that human talent, state sovereignty, and collective security transcend authoritarian dictates. Their existence demonstrates the self-defeating nature of Putin's imperial project and serves as a reminder that in international politics, the greatest threats often come not from our enemies' strength, but from our own strategic failures.

As Russia continues to pay the price for its leader's miscalculations, the question remains whether Putin will recognize the magnitude of his losses before they become irreversible, or whether his war

will be remembered as the moment when Russia chose isolation over
excellence, authoritarianism over prosperity, and imperial fantasy
over strategic reality. And an era where Russia gave away its future to
global china.

SECTION III
THE UKRAINIAN DYNAMIC

CHAPTER 1
MOVING BEYOND THE SOVIET MILITARY LEGACY

Ukraine's transformation from a Soviet-legacy defense system to a cutting-edge military innovator represents one of the most remarkable examples of adaptive mobilization in modern warfare. What began as an urgent response to Russia's 2014 annexation of Crimea evolved into a comprehensive restructuring that fundamentally reshaped not just Ukraine's military capabilities, but the very nature of 21st-century combat.

This transformation offers crucial lessons about institutional reform, technological innovation, and the integration of conventional and asymmetric warfare capabilities under existential pressure.

THE SOVIET LEGACY AND
THE CATALYST FOR CHANGE

When Russia invaded Ukrainian territory in 2014, Ukraine's defense industry remained overwhelmingly rooted in its Soviet past. The sector was exclusively state-owned and consolidated under Ukroboronprom, a massive conglomerate of Soviet-era enterprises characterized by bureaucratic inertia, corruption, and a culture of secrecy that significantly weakened military capabilities and deterred Western investment.[1]

In 2014, Ukraine possessed 134 enterprises within its military-industrial complex employing approximately 120,000 people, but the system was inefficient and focused on sustaining production of mature systems rather than developing new ones.

Russia's invasion of Crimea and the Donbas region created an urgent need to strengthen Ukraine's defensive capabilities. The government responded by steadily increasing military spending from a mere $62 million in military procurement in 2014 to $836 million by 2021, a thirteenfold increase.[2]

However, Ukraine's crucial advantage was recognizing that NATO deterrence might prove inadequate and beginning its modernization eight years before Russia's full-scale invasion. This head start would prove enormously valuable when broader conflict erupted in February 2022.

INSTITUTIONAL TRANSFORMATION: BREAKING THE SOVIET MOLD

The transformation of Ukroboronprom stands as perhaps the most significant institutional reform in Ukraine's defense sector. What had been a corrupt, inefficient Soviet-era concern began evolving into a more transparent, investor-friendly entity capable of leading large defense projects. Between 2014 and 2017, Ukroboronprom rose fourteen positions in the SIPRI world ranking of arms manufacturers, climbing from 91st to 77th place.

The reform process included several key elements. Multiple rounds of leadership changes brought fresh perspectives to state institutions and state-owned arms companies, culminating in the creation of a new Ministry of Strategic Industries in 2020. The transformation included converting Ukroboronprom from a state conglomerate into a joint-stock company, with plans to introduce OECD corporate governance standards across numerous state-run arms companies. New transparent arms procurement procedures replaced opaque Soviet-era practices that had facilitated corruption.[3]

However, these reforms faced significant challenges. According to then-Defense Minister Oleksiy Reznikov, only 13% of General Staff requests for newly developed or upgraded equipment were satisfied between 2014 and 2020. Roughly 20% of R&D projects in 2020 had been underway for twenty years or more with little to show for the investment with many serving primarily as vehicles for corruption.[4]

THE RISE OF PRIVATE INNOVATION

Problems in state-owned industries catalyzed the emergence of private arms companies that often surpassed their state-owned competitors in production efficiency, innovation, and adaptability. This represented a fundamental shift in Ukraine's defense ecosystem. In 2015, only 25% of state orders went to private companies. By 2020, this share had more than doubled to 54%, while state enterprises received just 36% and imports comprised the remaining 10%.[5]

Hundreds of small and medium enterprises entered the market, many focusing on cutting-edge capabilities like drones, advanced electronics, and AI-enabled systems. Before 2014, Ukraine lacked modern UAVs and relied on outdated Soviet models. After 2014, private organizations such as Athlon Avia and Aerorozvidka developed advanced systems that would prove crucial in the coming conflict.[6]

INTERNATIONAL PARTNERSHIPS: BREAKING RUSSIAN DEPENDENCE

A critical element of Ukraine's transformation involved deliberately shifting away from Russian supply chain dependence. Ukraine pursued joint ventures with countries like Turkey and Poland, yielding battlefield-leading systems like the Bayraktar TB2 drone. In February 2022, Ukraine signed a contract with Turkey for a $95.5 million factory near Kyiv to produce TB2 and TB3 drones, employing around 500 people including 300 Ukrainian engineers and technicians.[7]

Western partnerships also catalyzed crucial reforms, with partners urging Ukraine to address corruption and inefficiency in its arms companies. These relationships provided not just technology transfer but also institutional knowledge about modern defense industry practices that would prove invaluable when full-scale war arrived.

THE EMERGENCE OF HYBRID OPERATIONAL FORCE DESIGN

When Russia launched its full-scale invasion in February 2022, Ukraine's strategic response demonstrated a remarkable evolution in modern military strategy. At the core of Ukraine's defensive capability lies a dual foundation that represents a new paradigm in what can be termed hybrid operational force design or the seamless integration of conventional Western weaponry with indigenous technological innovations.

Western-supplied conventional weapons provided the essential backbone of Ukrainian military resistance, offering the firepower and reliability needed to sustain prolonged defensive operations. Systems like HIMARS, Javelin anti-tank missiles, and advanced air defense platforms proved their worth as the steady anchor of Ukrainian defense capabilities, providing the conventional military strength necessary to hold territory and repel advances.

However, it is Ukraine's own innovations, particularly in drone warfare, that have captured international attention and redefined modern battlefield tactics. Ukrainian forces pioneered new approaches to unmanned systems that extend far beyond traditional military doctrine, creating novel applications that have surprised military analysts worldwide. What sets Ukraine's approach apart is not simply the use of advanced Western weapons or innovative drone technology in isolation, but rather their ability to effectively combine these two elements.

This hybrid capability depends critically on effective command and control systems paired with advanced electronic warfare capabilities. These technological ecosystems serve as the nervous system that ties together disparate weapon systems, enabling coordinated operations that leverage both conventional firepower and innovative drone capabilities in synchronized attacks and defensive maneuvers.

Ukraine's strategic situation illustrates a fundamental principle of military doctrine: the mathematical disadvantage faced by defensive forces. Traditional military theory suggests that successful offensive operations require a three-to-one superiority in forces. Ukraine finds itself in the inverse position for they are "the one, not the three" facing numerically superior opposition while conducting defensive operations. This numerical disadvantage has necessitated the development of innovative tactics that multiply the effectiveness of available forces through superior tactical innovation and technological integration.

THE DRONE REVOLUTION: FROM ZERO TO GLOBAL LEADER

Ukraine's drone transformation exemplifies the broader shift to innovation at wartime speed. Starting from minimal capability in 2014, Ukraine became a world leader in drone technology and production. The evolution occurred in three distinct phases that fundamentally reshaped modern warfare.

Phase One in 2022 saw desperate adaptation of commercial DJI Mavic quadcopters and Chinese racing drones for military use, marking the beginning of Ukraine's shift from conventional military procurement to entrepreneurial innovation. Grassroots drone production emerged through volunteer networks like Social Drone UA, where civilians assembled drones in home workshops, garages, and even kitchens.[8]

Phase Two in 2023-2024 witnessed explosive scaling from 20,000 to 200,000 monthly drone production alongside development of indigenous long-range platforms like the UJ-22 Airborne. The government allocated $2 billion specifically for drone production in 2024, contracting 1.5 million drones in the first three quarters alone.[9]

Phase Three in 2025 achieved complete technological independence with 100% Ukrainian-made components and AI-integrated systems capable of autonomous targeting. Ukraine transformed from 100% import dependence to 96% domestic component sourcing within three years. Production capacity increased 4,500-fold from 2022 levels, while unit costs plummeted from millions to under $1,000 per drone. By 2025, Ukraine reached 200,000 drones per month, with annual production capacity exceeding 4.5 million units.[10]

Companies like Vyriy Drone achieved 100% localization using Ukrainian suppliers for motors, transmitters, and thermal cameras, while TAF scaled to 40,000 monthly production with $1 billion annual output. This transition from Chinese component dependence to complete domestic supply chains occurred in under three years, demonstrating unprecedented industrial mobilization under wartime conditions.[11]

OPERATION SPIDER WEB: THE CULMINATION OF INNOVATION

Operation Spider Web stands as the most sophisticated drone warfare operation in modern military history, representing the culmination of Ukraine's remarkable transformation from drone technology importer to global innovator. On June 1, 2025, Ukraine's Security Service executed a coordinated strike using 117 drones across five Russian airbases spanning 4,300 kilometers, destroying 41 aircraft worth $7 billion and demonstrating how asymmetric warfare can achieve strategic effects at unprecedented cost ratios.[12]

The operation required 18 months and 9 days of meticulous planning directly overseen by President Zelensky and executed by Security Service Chief Vasyl Maliuk. The operation's audacity lay not just in its geographic scope, spanning five time zones from Murmansk to

Siberia, but in its sophisticated integration of deception, technology, and strategic patience.[13]

Ukrainian operatives established a command center directly adjacent to an FSB regional headquarters in Russia, working undetected for months while positioning assets across the continent. The "Trojan horse" strategy concealed 117 Ukrainian-made Osa quadcopters inside wooden cabins mounted on commercial trucks, with remotely-operated roofs that deployed the drones on command. Unwitting Russian drivers transported these systems to positions near five strategic airbases, never knowing they carried Ukraine's most sophisticated weapons.[14]

The execution demonstrated unprecedented technical sophistication. Each drone carried a 3.2-kilogram payload and operated through Russian 4G/LTE mobile networks, controlled by individual operators working from Ukraine thousands of kilometers away. AI-assisted targeting systems trained on Soviet aircraft displayed in Ukraine's Poltava Museum enabled drones to identify vulnerable points on aircraft with 90-centimeter precision. The system combined ArduPilot open-source autopilot software with autonomous navigation capabilities, allowing drones to complete missions even when communication links were severed.[15]

The breakthrough AI targeting system exemplifies Ukrainian innovation philosophy. Rather than developing expensive military-specific software, Ukrainian engineers trained artificial intelligence using museum aircraft from their own Poltava Museum of Long-Range Aviation. This approach created precise targeting algorithms capable of identifying critical components, fuel tanks, missile pylons, wing roots, while costing orders of magnitude less than traditional military systems.[16]

Operation Spider Web demonstrated $600-1,000 FPV drones destroying $250 million bombers, creating unsustainable exchange ratios for defenders. By destroying approximately one-third of Russia's strategic bomber fleet, Ukraine eliminated platforms responsible for cruise missile attacks on Ukrainian cities while demonstrating that geographic distance no longer provided security for strategic assets.

INTERNATIONAL INTELLIGENCE COOPERATION: THE SPACE DIMENSION

The ongoing conflict fundamentally transformed international cooperation in space-based intelligence, with Finland and Eastern European nations playing pivotal roles in establishing unprecedented satellite data-sharing arrangements. These partnerships represent a significant shift in how democratic nations approach military intelligence sharing and highlight the critical importance of commercial space capabilities in modern warfare.

At the forefront of this cooperation stands Finland's ICEYE, a commercial satellite company that became Ukraine's most significant space-based intelligence partner. In August 2022, ICEYE signed a groundbreaking contract with the Serhiy Prytula Charity Foundation, providing Ukraine with exclusive control over one of its Synthetic Aperture Radar (SAR) satellites while granting access to the company's entire constellation.[17]

This partnership proved remarkably effective. According to a 2024 Newsweek report, nearly two-fifths of the information gathered from Finland's ICEYE satellites was used to directly prepare strikes against Russian forces, resulting in damage estimated in the billions of dollars. ICEYE's SAR satellites possess unique capabilities that traditional optical satellites cannot match. They can capture high-resolution images during darkness, through cloud cover, and in all weather conditions. The technology can reveal heat signatures from running engines, detect camouflaged vehicles, and map terrain changes caused by artillery strikes.[18]

The success of bilateral partnerships catalyzed a much broader international initiative. In February 2023, eighteen Western nations signed a letter of intent establishing the Allied Persistent Surveillance from Space Initiative (APSS), a NATO-supported program designed to revolutionize intelligence sharing among democratic allies. The

APSS coalition includes key Eastern European nations such as Belgium, Bulgaria, Hungary, and Romania, alongside Finland, France, Italy, the Netherlands, Norway, Portugal, Spain, Turkey, Sweden, the United Kingdom, United States, and Canada.[19]

The program, which became operational in 2025 under the codename "Aquila," focuses on three primary objectives: sharing data from national surveillance satellites, jointly processing and analyzing intelligence data, and collectively funding the purchase of commercial satellite imagery and intelligence products. Participating nations committed a total of $1 billion to the initiative, with Luxembourg alone contributing €16.5 million.[20]

In April 2025, Japan made a historic decision to share synthetic aperture radar satellite images from its Institute for Q-shu Pioneers of Space (iQPS) with Ukraine's military intelligence agency. This marked the first time Japan has ever shared such sensitive geospatial intelligence with any foreign nation, representing a significant technological leap as iQPS operates some of the world's most advanced lightweight SAR satellites.[21]

The Ukraine conflict demonstrated how commercial satellite companies can serve as force multipliers in military operations. Unlike traditional government-operated intelligence satellites, commercial providers offer rapid deployment, frequent revisit capabilities, and the flexibility to adapt quickly to changing battlefield requirements. ICEYE operates the world's largest fleet of commercial SAR satellites with 21 spacecraft successfully launched to date, delivering near-real-time intelligence invaluable for Ukrainian forces planning operations, tracking Russian logistics routes, and identifying high-value targets.

THE STARLINK IMPACT

Starlink, the satellite internet service developed by Elon Musk's SpaceX, has played a pivotal role in Ukraine's defense and resilience since the beginning of Russia's full-scale invasion in February 2022.[22]

By providing reliable, high-speed internet access even in war-torn and remote areas, Starlink terminals have become a lifeline for both the Ukrainian military and civilians, enabling communication, coordination, and morale-boosting engagement across the country.

The Ukrainian military quickly recognized the strategic value of Starlink. Traditional communication infrastructure, including cell towers and fiber-optic cables, was vulnerable to Russian attacks and electronic warfare. Starlink's low-latency, secure satellite internet allowed Ukrainian forces to maintain command and control, coordinate troop movements, and relay intelligence in real time—even in regions where terrestrial networks were destroyed or jammed.

Starlink terminals were deployed at the front lines, in command centers, and on military vehicles, providing a resilient network that was difficult for Russia to disrupt. This capability proved essential for battlefield management, drone operations, and cyber defense, giving Ukraine a critical advantage in situational awareness and rapid response.

Beyond the military, Starlink empowered Ukraine's civilian population to actively support the war effort. In cities and villages under siege or cut off from conventional internet, Starlink enabled residents to stay informed, share information, and participate in volunteer initiatives. Social media and messaging platforms, accessible via Starlink, became tools for organizing humanitarian aid, documenting war crimes, and boosting national morale.

The service also allowed Ukrainian journalists and activists to report from conflict zones, sharing real-time updates with the world. This transparency helped galvanize international support and keep global attention focused on Ukraine's struggle.[23]

Starlink's availability inspired grassroots mobilization. Civilians used the internet to coordinate volunteer efforts, from delivering supplies to evacuating the wounded. Crowdfunding campaigns for drones, medical equipment, and other military needs flourished online, with Starlink ensuring that even remote communities could contribute financially and logistically.

The Ukrainian government and NGOs leveraged Starlink to maintain contact with displaced persons and refugees, ensuring that humanitarian aid reached those most in need. This connectivity helped preserve a sense of unity and purpose among Ukrainians, even as the war displaced millions.

The psychological impact of Starlink cannot be overstated. In the face of relentless attacks and information blackouts, access to the internet gave Ukrainians a sense of normalcy and hope. Families could stay in touch, students could continue their education, and citizens could access entertainment and news, reinforcing national resilience and determination.

Starlink also played a role in countering Russian propaganda by providing reliable sources of information. Ukrainians could verify facts, share their own stories, and challenge misinformation, strengthening their collective resolve to defend their country.

Despite its benefits, Starlink has not been without challenges. Early in the war, there were concerns about service interruptions and the potential for terminals to be targeted by Russian electronic warfare. SpaceX and Ukrainian authorities worked to address these issues, but the reliance on a single commercial provider has raised questions about long-term sustainability and strategic independence.[24]

Additionally, the cost and logistics of distributing and maintaining terminals remain significant, especially in the most dangerous frontline areas. However, ongoing international support and local ingenuity have helped Ukraine overcome many of these obstacles.

Starlink's impact on Ukraine's war effort has been transformative. By providing secure, resilient internet access, it has empowered the military to operate effectively and enabled civilians to participate actively in the defense of their country.[25] The service has not only strengthened Ukraine's ability to resist Russian aggression but has also fostered a sense of unity and resilience among its people.

THE UKRAINIAN DEFENSE PRODUCTION ECOSYSTEM TODAY

Ukraine's defense transformation yielded impressive economic results. By 2023, Ukrainian Defense Industry (the rebranded Ukroboronprom) achieved a 69% year-on-year increase in arms revenues to $2.2 billion, the fastest increase and highest revenues the company had ever recorded. This made it the fastest-growing company among the world's top 100 defense firms, rising from 65th to 49th place in global rankings.[26]

The broader defense ecosystem flourished as well. Ukraine's defense industry now employs about 67,000 highly skilled workers and encompasses hundreds of companies ranging from major state enterprises to small volunteer workshops.[27]

Beyond drones, Ukraine achieved mass production across multiple weapons categories. The 2S22 Bohdana howitzer went from a single prototype in 2022 to over 184 units produced by 2024. Artillery shell production resumed across multiple calibers, small-arms ammunition production restarted for the first time since 2014, and approximately 100 R-360 Neptune cruise missiles were produced in 2024 alone.[28]

Ukraine developed everything from small FPV reconnaissance drones to sophisticated long-range attack systems like the Peklo, capable of striking targets 700 kilometers away. This comprehensive production capability created what essentially amounts to a private-sector national geospatial and defense agency, allowing Ukraine to leverage dedicated capabilities without having developed its own space launch program or traditional defense industrial base, a model that could prove attractive to other nations facing security challenges.

CHALLENGES AND ONGOING CONSTRAINTS

Despite remarkable achievements, Ukraine's transformation faced significant constraints. The biggest barrier to expansion remained financial. Ukraine's government and private sector were producing more drones and weapons than the state could afford to acquire. Foreign investment and military aid remained crucial for sustaining operations. Ukraine still cannot produce equipment to intercept hypersonic and ballistic missiles, highlighting ongoing technological gaps.

Russian missile strikes on Ukrainian facilities forced much production to remain in small, distributed workshops rather than large, efficient factories. Dependence on Chinese components persisted despite efforts at localization. While companies achieved domestic production of specialized components like drone motors, dependence continued for basic materials like magnets and advanced manufacturing equipment.

Russia has alleged that over 200 commercial satellites and 20 analytical centers from Western countries are providing intelligence products to Ukraine, indicating Moscow's awareness of these capabilities and potential development of countermeasures. Finland has already experienced increased satellite jamming since 2022, believed to originate from Russia, which has disrupted both civilian and military satellite services.[29]

In short, Ukraine's transformation from a Soviet-legacy defense system to a modern, innovative military-industrial powerhouse represents one of the most remarkable examples of adaptive mobilization in contemporary warfare.

Hybrid operational force design maximizes defensive capabilities. The seamless integration of conventional Western weaponry with indigenous technological innovations, tied together through sophisticated command and control systems, created a force multiplier effect that allowed Ukraine to overcome traditional numerical disadvantages through superior tactical innovation and technological integration.

CHAPTER 2

THE DIGITAL ARSENAL:
THE ISRAELI-UKRAINIAN
PARTNERSHIP

T he remarkable effectiveness of Ukrainian drone warfare in repelling Russian aggression has captured the attention of defense analysts worldwide.

While much attention has focused on Ukrainian innovation since the 2022 full-scale invasion, a crucial yet under examined factor lies in Ukraine's pre-war technology partnerships, particularly with Israel.

The foundation for Ukraine's current drone supremacy was laid years before the first Russian missile crossed the border, built on a sophisticated collaboration in defense technology, digital engineering, and software-centric warfare that fundamentally transformed how Ukrainian forces think about unmanned systems.

This partnership represents more than simple arms sales or technology transfer. It reflects a profound shift in defense thinking: from platform-centric to network-centric warfare, from hardware perfection to software adaptability, and from lengthy development cycles to rapid iteration. Understanding this relationship is essential for comprehending not just Ukraine's current battlefield success, but the future of autonomous warfare itself.

THE FOUNDATION: PRE-WAR UKRAINIAN-ISRAELI DEFENSE COOPERATION

Ukraine's relationship with Israel in the defense sector began long before 2022. Israeli-designed drones and Israeli-assisted designs became early components of Ukraine's arsenal well before the full-scale war.[1] This cooperation extended beyond simple hardware procurement to encompass electronic warfare equipment, counter-drone technologies, and critically, the software architectures that enable rapid adaptation of unmanned systems.

Israel has maintained its position as a global leader in autonomous systems, particularly in drone and counter-drone capabilities, from tactical UAVs to loitering munitions like the Harop. What distinguished Israeli defense innovation was its software-centric approach that prioritized adaptable software architectures over hardware perfection.[2] This philosophy proved directly applicable to rapid drone iteration in combat conditions, a reality that would define Ukraine's wartime innovation.

Ukrainian defense firms maintained extensive collaboration with Israeli companies on electronic warfare, UAV technology, and command, control, communications, computers, intelligence, surveillance, and reconnaissance (C4ISR) systems.

This wasn't merely about purchasing equipment. It represented genuine technology transfer and joint development work.[3]Ukrainian engineers who worked with Israeli firms absorbed methodologies for distributed systems, agile software development, and the integration of sensors and shooters, exactly the capabilities required for effective "kill webs" rather than traditional "kill chains."

UKRAINE'S DIGITAL FOUNDATION: A SOFTWARE POWERHOUSE

To understand why Ukraine could so effectively leverage Israeli expertise, one must recognize Ukraine's pre-war position as an IT powerhouse. Before February 2022, Ukraine's IT sector had established itself as a major exporter of IT services within Europe, with annual growth rates between 25-30%.[4] By 2021, the IT industry generated $6.8 billion in exports, a 36% increase from the previous year—and employed approximately 285,000 specialists.[5]

Ukraine ranked 8th globally in cutting-edge technical knowledge according to the 2022 Coursera Global Skills Report.[6] This wasn't merely about quantity. Ukrainian developers were recognized for technical expertise in software development, testing and quality assurance, and IT consulting. The country produced nearly 25,000 tech graduates annually from universities that appeared in QS World University Rankings. Ukrainian IT companies offered comprehensive services ranging from software development to cybersecurity, primarily serving clients in the United States, Canada, and Western Europe.

This digital infrastructure expertise, networking, distributed systems, agile software development, created a foundation directly applicable to drone swarm coordination, electronic warfare integration, and sophisticated targeting capabilities. The Ukrainian government's Diia City initiative, launched just weeks before the 2022 invasion, created a special legal and tax framework designed to incentivize tech innovation and attract foreign investment.[7] By 2024, over 1,560 companies had become residents of this framework, including prominent international players like Samsung, Revolut, and Lyft.

THE SYNTHESIS: SOFTWARE SKILLS MEET DEFENSE REQUIREMENTS

When Russia launched its full-scale invasion, Ukraine's combination of Israeli-influenced defense thinking and domestic software engineering prowess proved decisive. Ukraine rapidly developed an ecosystem capable of producing approximately 2 million drones in 2024, with over 200 Ukrainian companies now operating in the UAV sector.[8] More remarkably, Ukraine introduced more than 200 domestically developed unmanned aerial systems and over 40 ground robotic platforms since February 2022, with most entering service in 2024 alone.

Ukrainian engineers leveraged their software expertise to create systems using consumer-grade components, open-source software, and 3D printing. This approach compressed development cycles that might take years in traditional aerospace manufacturing into weeks. As one defense analyst noted, Ukraine's rapid iteration process allows prototypes that work today to be adjusted by tomorrow, with parts printed locally or sourced from regional suppliers.[9] Ukrainian developers collaborate through informal networks, sharing code and hardware solutions to overcome challenges like electronic interference and GPS jamming.

This capability for rapid software iteration proved critical. Jack Watling of the Royal United Services Institute observed that Ukraine, using specialists from its tech sector, has been able to make software adaptations to keep combat UAVs effective in as little as two weeks.[10] Ukrainian airframes allow "rapid insertion of modular software change," and if software isn't updated within six to twelve weeks, adversaries gather sufficient data to develop effective countermeasures. This compressed timeline, fundamentally a software problem rather than a hardware problem, plays directly to Ukraine's strengths.

Ukrainian engineers increasingly leverage open-source technologies and existing computer vision models to accelerate research and development while keeping costs low. By integrating readily available software solutions, including open-source computer vision frameworks like YOLOv8, developers can rapidly create and deploy autonomous functions.[11] Ukraine's massive drone attack that destroyed a significant portion of Russia's strategic bomber fleet in 2025 utilized ArduPilot, a nearly 20-year-old open-source drone autopilot software, demonstrating how software expertise can transform commercial-grade systems into strategic weapons.[12]

THE COGNITIVE ADVANTAGE:
FROM PLATFORMS TO NETWORKS

The Ukrainian approach to drone warfare reflects a fundamental shift from platform-centric to kill web operations precisely the evolution that Israeli defense thinking had pioneered. Ukraine isn't winning the drone war because it has better hardware; in many cases, it doesn't. Ukraine wins because it can write better code, iterate faster, and integrate sensors and shooters more effectively. This represents a cognitive advantage enabled by the digital engineering culture that Israeli partnership reinforced.

Recent battlefield developments demonstrate this advantage clearly. AI-enabled autonomous navigation capabilities are driving marked decreases in overall strike costs by minimizing both drone losses and repeated mission attempts. Systems can often achieve objectives using just one or two drones per target rather than eight or nine, and fully autonomous flight makes systems reusable. Preliminary data from one brigade showed hit rates improving from 20% to 80%, a fourfold increase achieved through enhanced target tracking capabilities rather than hardware improvements.[13]

The Israeli influence is evident in Ukraine's emphasis on distributed operations and electronic warfare resilience. Companies like Zvook developed AI-powered acoustic detection systems that can detect drones from several kilometers away with 97.5% accuracy, technology now attracting significant interest from Israeli defense firms facing similar Iranian drone threats.[14]This reversal of technology flow, with Israel now seeking Ukrainian expertise, validates the effectiveness of Ukraine's software-centric approach to warfare.

THE REVERSAL: ISRAEL
LEARNING FROM UKRAINE

The relationship between Ukrainian and Israeli defense innovation has evolved dramatically since October 2023. Israel, facing escalating drone attacks from Iran and its proxies with approximately 1,300 drone strikes since October 7, 2023 is now actively seeking Ukrainian expertise.16 Ukrainian Ambassador to Israel Yevhen Korniychuk confirmed growing interest from Israeli firms such as Elbit and R2 in Ukrainian solutions, noting that "Israeli private companies and state defense firms are quite actively communicating with their Ukrainian counterparts right now."[15]

This represents a remarkable evolution. Before the Gaza war, Israeli officials largely ignored Ukrainian warnings about Iranian drone threats and offers to collaborate. A former senior Ukrainian official revealed that Ukrainian companies developing anti-drone technology sent representatives to Israel before October 7, but "could not find a single counterpart from the defense sector in Israel willing to meet with them."[16]

The situation has transformed dramatically. Israel now recognizes that Ukraine's battlefield-tested innovations developed under the pressure of facing over 7,000 Russian drone attacks in 2024 alone offer solutions unavailable elsewhere.

Defense analyst Ian Matveev stated bluntly: "When it comes to working with UAVs and countering them, the Ukrainian army is currently number one in the world."[17]

An Israeli military source acknowledged that Ukraine's practical battlefield experience offers Israel valuable knowledge in counter-drone operations. Israeli companies are now studying Ukrainian production techniques, particularly distributed assembly lines and reliance on additive manufacturing, recognizing that Ukraine's approach to rapid prototyping challenges established manufacturing models.

STRATEGIC IMPLICATIONS: THE FUTURE OF AUTONOMOUS WARFARE

Ukraine's drone warfare success reveals fundamental truths about modern conflict. The shift from platform-centric to network-centric warfare prioritizes software adaptability over hardware perfection, rapid iteration over lengthy development cycles, and distributed intelligence over centralized control. These principles deeply embedded in both Israeli defense thinking and Ukrainian software engineering culture represent the future of military operations.

The Ukrainian-Israeli technology partnership before 2022 was indeed consequential, but not in the conventional sense of arms transfers or equipment sales. It was consequential because it exposed Ukrainian engineers and military planners to a fundamentally different way of thinking about warfare, one where software determines outcomes, where networks matter more than platforms, and where the ability to iterate rapidly trumps initial perfection.

Ukraine absorbed these lessons and, combining them with its own formidable software engineering capabilities, created an approach to autonomous warfare that now serves as a model for military forces worldwide.

As General Patton observed, "if everyone is thinking alike, someone isn't thinking." This is the motto on our *Second Line of Defense* website from the beginning.

Ukraine's pre-war investment in Israeli-influenced digital infrastructure, its willingness to think differently about warfare, may prove one of the war's most significant strategic factors.

The cognitive advantage this partnership enabled continues to shape battlefield outcomes, demonstrating that in modern warfare, superior code may matter more than superior hardware. This represents not just a tactical evolution, but a fundamental transformation in how nations must prepare for conflict in the 21st century.

CHAPTER 3
WAR IN THE MODERN AGE

T he war in Ukraine is the most documented conflict in human history, and at the center of this digital chronicle sits Telegram, a messaging application that has transformed from a simple communication tool into a multifaceted instrument of warfare.

Since Russia's full-scale invasion began in February 2022, Telegram has served simultaneously as a lifeline for civilians, a command-and-control platform for military operations, a propaganda megaphone, and a recruitment tool for espionage.

Understanding Telegram's role in this conflict reveals how modern warfare extends far beyond traditional battlefields into the digital realm where information itself becomes both weapon and shield.

THE RISE OF TELEGRAM AS UKRAINE'S PRIMARY INFORMATION SOURCE

Before the invasion, Telegram was already popular in Eastern Europe, but the war catapulted it to unprecedented prominence. By 2023, an astonishing 72% of Ukrainians cited Telegram as a primary news source, representing a significant increase from 63% the previous year.[1] Its rise occurred as traditional social media platforms faced restrictions and censorship, particularly in Russia, where Facebook, Instagram, and Twitter were either banned or severely limited by authorities.[2]

The platform's popularity stems from its unique combination of features: end-to-end encrypted private messaging, public broadcasting channels capable of reaching millions, and minimal content moderation that allows for the rapid, unfiltered dissemination of information. For Ukrainians, Telegram became the most reliable way to receive real-time updates about air raids, safe routes, available bomb shelters, and the general security situation.

For refugees fleeing the violence, Telegram proved essential in coordinating the exodus of more than three million people. The app connected displaced Ukrainians to safe routes, humanitarian aid, and shelter information across Europe. Millions of Ukrainians living abroad used the platform to maintain connections with their homeland, desperately scanning feeds for familiar landmarks or faces that might provide news of loved ones.[3]

A TOOL FOR CIVILIAN SAFETY AND GOVERNMENT COMMUNICATION

Ukrainian authorities swiftly recognized Telegram as a vital pillar of wartime communications infrastructure. Throughout the country, official Telegram channels run by local governments and agencies became essential tools for disseminating urgent safety information, including real-time air-raid warnings, bomb shelter locations, security instructions, and alerts on identifying potential Russian saboteurs.[4]

Telegram's two-way design uniquely empowered Ukrainian civil defense. While officials broadcast critical updates to mass audiences, citizens contributed intelligence by reporting details such as enemy troop movements, armored vehicles, and suspicious activities via dedicated Telegram bots and chat platforms. This real-time crowdsourced data proved invaluable, Ukraine's security services credited information received through these chatbots for enabling a successful strike against Russian vehicles near Kyiv in March 2022, with officials noting that user submissions were responsible for "new trophies every day"[5]

This crowdsourced intelligence gathering represented a novel approach to civil defense, turning ordinary citizens into the eyes and ears of the military establishment. The democratization of intelligence collection through Telegram fundamentally altered the traditional relationship between civilians and military forces during wartime.

INTELLIGENCE SERVICES
ENTER THE PUBLIC SPHERE

Perhaps one of the most unprecedented aspects of Telegram's role in the war has been its adoption by Ukrainian intelligence agencies as a public-facing communication platform. Ukraine's Main Directorate of Intelligence (HUR) developed a structured communication strategy on Telegram that diverges sharply from conventional intelligence practice, which typically emphasizes secrecy and limited public disclosure.

The verified Ukrainian-language Telegram channel operated by the Main Directorate of Intelligence (HUR) accumulated more than 240,000 followers by February 2024. In-depth analysis of 2,606 messages posted from the onset of Russia's full-scale invasion up to February 2024 identified three core functions: establishing institutional legitimacy, targeting adversaries through carefully calibrated disclosures, and mobilizing domestic citizens for support and participation. The HUR's channel delivered consistent battlefield updates, released operational footage, publicized intercepted Russian communications, announced special intelligence projects, and broadcast calls to civilians for supportive action.[6]

This approach transformed intelligence communication from episodic outreach into an ongoing process of public engagement.

Rather than maintaining the traditional veil of secrecy, Ukrainian intelligence services used digital platforms to coordinate narrative control, reinforce their legitimacy, and enlist the public as active contributors to the intelligence mission.

This represents a fundamental shift in how intelligence agencies operate during high-intensity conflicts, suggesting that transparency and public engagement can complement rather than compromise operational effectiveness.

MILITARY UNITS AND FRONTLINE DOCUMENTATION

Beyond government agencies, individual Ukrainian military units established their own Telegram channels to share updates, boost morale, and document their operations. Elite units like the Third Separate Assault Brigade, the 36th Separate Marine Brigade, and the 47th Separate Mechanized Brigade maintained active channels showcasing training exercises, battlefield successes, and appeals for equipment donations.[7]

These military channels served multiple purposes. They provided transparency to Ukrainian citizens about how their armed forces were performing, helped maintain public support for the war effort, and attracted international attention and assistance. The channels also documented the war in unprecedented detail, with soldiers posting footage from helmet cameras, drones, and battlefield positions.

Military bloggers, known as "voenkory" in Russian and Ukrainian, emerged as influential voices on Telegram. These individuals, often with military experience or embedded with combat units, provided detailed tactical analysis and frontline reports that sometimes offered greater technical detail than official media sources.[8] The phenomenon of voenkory represented a new category of war correspondents, semi-official voices that blurred the lines between journalism, propaganda, and military communication.

PROPAGANDA AND PSYCHOLOGICAL WARFARE

While Telegram served beneficial purposes for Ukraine, it also became a powerful tool for Russian propaganda and psychological warfare. The unregulated nature of the platform made it ideal for disseminating disturbing content designed to demoralize Ukrainian forces and terrorize civilians.

Russian channels regularly posted graphic footage of killed Ukrainian soldiers, often accompanied by mocking captions and propaganda narratives. One particularly notorious channel, "Arkhangel Spetsnaza," grew to nearly 1.2 million subscribers by monetizing "exclusive" violent content. The channel offered subscription tiers costing between 5 and 10 dollars monthly, granting access to graphic videos of Russian military operations. Investigations revealed that during just two months, this channel collected approximately 30 million rubles (about $317,000) in funding, which was purportedly used to purchase drones for Russian forces.[9]

The psychological impact of such content extends beyond the immediate shock value. By flooding Telegram with graphic imagery and propagandistic narratives, Russian sources sought to create a sense of Ukrainian hopelessness and inevitability about Russian victory. This "war porn," as some analysts termed it, served both to rally pro-Russian audiences and to demoralize Ukrainian supporters.[10]

COORDINATED DISINFORMATION CAMPAIGNS

Beyond individual channels, Russian operatives conducted sophisticated, coordinated disinformation campaigns on Telegram. Research revealed that from January 2024 to April 2025, a network of 3,634 inauthentic Telegram accounts posted more than 316,000 pro-Russian comments across channels focused on Ukraine's temporarily occupied territories.[11] These comments promoted pro-Russian propaganda, anti-Ukrainian narratives, and abstract calls for peace, often carefully tailored to local conditions and current events.

This massive bot network represented Russia's attempt to extend its occupation into digital spaces, shaping the information environment in territories under its control. The sophisticated nature of the operation with thousands of accounts posting hundreds of thousands of coordinated messages demonstrated the strategic importance Russia placed on controlling the Telegram information landscape.

The Internet Research Agency (IRA), a Kremlin-affiliated entity known for conducting online influence operations, was at the forefront of Russia's coordinated efforts to manipulate information environments in support of state interests. In February 2023, Yevgeny Prigozhin, leader of the Wagner mercenary group and a close associate of President Vladimir Putin, openly acknowledged that he had founded and managed the IRA, issuing his statement via a Wagner-affiliated Telegram channel. Prigozhin further explained that the agency was established to defend Russia's information space against Western propaganda, marking a rare explicit admission of such high-level Kremlin-linked operations.[12]

This admission confirmed long-standing suspicions about state-coordinated manipulation of social media platforms, including Telegram.

Russian state-sponsored youth influencers also leveraged Telegram to disseminate patriotic messaging to young Russians. These influencers used social media techniques to normalize official stances on the war, creating the impression of grassroots support while actually amplifying state-approved narratives. Their efforts focused on defining "good patriots" as those who participated in memory politics, engaged in militarized activities, and supported the war effort, whether physically or virtually.[13]

ESPIONAGE AND RECRUITMENT

Telegram's role extended even to espionage recruitment. For at least six months, pro-war Telegram channels actively recruited Russian-speaking residents of Europe to spy on NATO military sites and report findings through specialized bots. These channels distributed instructions on photographing military bases, purchasing local maps and guidebooks, and using local SIM cards to avoid detection.[14]

Recruitment messages distributed on Telegram often originating in smaller groups but rapidly amplified through wider networks have reached tens of thousands of potential recruits across Europe. European security services have documented numerous cases in which individuals were recruited via Telegram to carry out surveillance, vandalism, and arson attacks targeting military infrastructure and critical sites.

Authorities in Latvia, Germany, and Poland have arrested suspects allegedly enlisted through Telegram to conduct sabotage operations in support of Russian interests, with cases including attempted arson, attacks on military facilities, and the disruption of supply lines. These incidents illustrate a pattern of decentralized, low-level operations enabled by encrypted platforms, posing new challenges to European security services.[15]

This weaponization of Telegram for intelligence recruitment represented a democratization of espionage. Rather than relying on traditional, labor-intensive recruitment methods, Russian intelligence services could cast a wide net through Telegram, identifying and activating sympathizers across Europe with minimal risk to handlers.

CONCLUSION

Telegram's multifaceted role in the Russia-Ukraine war reveals how modern conflicts are waged simultaneously in physical and digital realms. The platform has served as a lifeline for civilians seeking safety information, a command-and-control network for military operations, a public relations tool for intelligence agencies, a documentation archive for war crimes, and a weapon of propaganda and psychological warfare.

The same technological features, encryption, broad reach,

minimal moderation serve both humanitarian and malicious purposes. Ukrainian civilians use Telegram to find bomb shelters and report Russian troop movements; Russian operatives use it to spread disinformation and recruit spies. This duality encapsulates the fundamental challenge of digital communication in wartime: the tools that empower also endanger.

As conflicts increasingly extend into digital spaces, the Ukraine war offers crucial lessons about the role of social media platforms in modern warfare. The unprecedented documentation of military operations, the use of crowdsourced intelligence, the public-facing strategies of intelligence agencies, and the sophisticated disinformation campaigns all point toward a future where information warfare is as critical as kinetic warfare.

Telegram's role in Ukraine demonstrates that winning the digital battle, controlling narratives, maintaining morale, documenting atrocities, and countering disinformation, is inseparable from success on the physical battlefield.

CHAPTER 4

UKRAINE, DEMOCRATIC DEFICITS, AND STRATEGIC NECESSITY

W estern policy toward Ukraine confronts a fundamental tension: Supporting a nation with significant democratic deficits and endemic corruption while framing the conflict as a struggle between democracy and authoritarianism.

Ukraine ranks as "Partly Free" in Freedom House assessments, with a score of 61 out of 100, placing it below the threshold for consolidated democracies.[1]

Transparency International's 2023 Corruption Perceptions Index ranks Ukraine 104th out of 180 countries, indicating systemic corruption challenges that persist even amid war.[2]

These realities complicate the narrative of defending democratic values while raising legitimate questions about resource allocation, moral consistency, and strategic coherence.

The dilemma is neither new nor unique to Ukraine. Throughout the Cold War and its aftermath, the United States and its allies maintained partnerships with deeply flawed governments when strategic interests aligned, from authoritarian regimes in East Asia to corrupt post-colonial states in Africa and the Middle East.

What distinguishes the Ukrainian case is the explicit framing of support as a defense of democratic principles, creating a rhetorical commitment that invites scrutiny when reality falls short.

This chapter examines multiple frameworks for understanding these tensions, arguing that while Ukraine's governance problems are real and significant, they do not negate the strategic case for support, though they do require honest acknowledgment and a more nuanced policy rationale than simple democratic solidarity.

THE IMPERFECT INSTRUMENT PROBLEM

Supporting Ukraine does not require pretending it resembles Denmark or Switzerland. The analytically useful question is not "Is Ukraine a model democracy?" but rather "What are the strategic consequences of Russian success or failure in Ukraine?"

This reframing moves the debate from normative assessment to strategic calculation. As historian Michael Kimmage argues, "Ukraine's imperfections do not erase the fact that Russian victory would fundamentally alter the European security order and embolden revisionist powers globally."[3]

The choice facing Western policymakers is not between perfect and imperfect allies but between available options with dramatically different strategic trajectories.

A Russian military victory would establish several dangerous precedents.

- First, it would demonstrate that conventional military conquest remains viable in the 21st century, potentially inspiring other revisionist powers with territorial grievances.
- Second, it would signal that nuclear-armed states can successfully coerce non-nuclear neighbors without meaningful Western intervention, undermining nonproliferation frameworks that depend on security assurances.

- Third, it would shift the European security architecture away from the post-1945 principle that borders should not be changed by force, potentially destabilizing regions from the Balkans to the Caucasus.

Moreover, Russian success would have cascading effects on NATO credibility and alliance cohesion. If the United States and European powers prove unwilling or unable to prevent the conquest of a country seeking Western integration, allies in Eastern Europe would reasonably question the value of collective security commitments. This could drive countries like Poland, the Baltic states, and Romania toward independent nuclear deterrents or accommodationist policies toward Moscow, outcomes that would fragment the transatlantic security community and increase rather than decrease the risk of future conflicts.

From this strategic perspective, Ukrainian governance quality becomes a secondary consideration. The primary question is whether preventing Russian conquest serves Western interests by maintaining deterrence credibility, preserving the norm against aggressive war, and containing revisionist powers. Ukraine's imperfections do not alter this strategic calculus. They simply make the policy more difficult to sustain politically and require more honest public communication about what is actually being defended.

WARTIME GOVERNANCE
VERSUS PEACETIME STANDARDS

Assessing Ukraine's democratic credentials during existential war involves a category error: judging wartime governance by peacetime standards. Since Russia's 2022 invasion, President Volodymyr Zelenskyy has consolidated power, postponed elections, restricted certain opposition parties (particularly those with alleged ties to Russia), and imposed controls on media coverage. In peacetime, such measures would signal authoritarian drift and merit serious concern. During a war for national survival, however, they may reflect legitimate security imperatives rather than systematic democratic backsliding.

Ukraine's Constitution explicitly provides for suspension of certain rights during martial law, a provision found in most democratic constitutions including those of established Western democracies.[4]

The United States suspended habeas corpus during the Civil War, interned Japanese Americans during World War II, and expanded executive powers dramatically after September 11, 2001. France has repeatedly invoked emergency powers to address terrorist threats. Israel has maintained military government over disputed territories for decades while claiming democratic legitimacy. These precedents do not justify all wartime measures, but they establish that democratic governance exists on a spectrum that shifts based on security circumstances.

The critical test for Ukrainian democracy will come in the postwar transition.

- Does the Zelenskyy government relinquish emergency powers when hostilities end?
- Does Ukraine return to competitive elections with genuine opposition participation?
- Do independent media and civil society organizations regain space to criticize government policy?

These questions cannot be answered definitively while the war continues, creating irreducible uncertainty about Ukraine's democratic trajectory.

Historical precedents offer mixed guidance. Some countries emerged from existential conflicts with strengthened democratic institutions, France after World War II, Poland after 1989. Others saw wartime centralization become permanent, Turkey after its founding conflicts, Egypt after the wars with Israel.

The outcome depends partly on institutional design, partly on political culture, and partly on post-war economic conditions and external pressure. Western support creates leverage for encouraging democratic restoration, but only if exercised consistently and backed by concrete conditionality.

THE TRANSFORMATION ARGUMENT

Ukraine's dramatic military evolution since 2014 demonstrates institutional capacity for adaptation under pressure. The Ukrainian Armed Forces transformed from a dysfunctional post-Soviet military plagued by corruption and low morale into an effective fighting force capable of defending against a numerically superior adversary.

This transformation involved not just weapons acquisition but fundamental changes in military culture, training methodologies, and operational doctrine. If Ukraine can reform its military under battlefield conditions, the argument goes, perhaps governance reform is equally achievable.

This optimistic assessment must be qualified. Military reform succeeded partly because it had clear metrics for success (battlefield effectiveness), concentrated decision-making authority, and direct Western mentorship with conditionality attached to aid.

Governance reform operates differently: success metrics are contested, authority is diffuse across multiple institutions, and external actors have less direct leverage over domestic political processes. Corruption involves complex networks of political, economic, and criminal interests that are far more difficult to dismantle than outdated military doctrines.

Moreover, wartime conditions can actually entrench corruption rather than reduce it. The enormous flows of international aid, the suspension of normal procurement procedures, and the emergency atmosphere all create opportunities for graft. Some Ukrainian officials and private actors have exploited these conditions, as documented by Ukrainian investigative journalists and anti-corruption organizations. The question is whether post-war reconstruction will create pressure for reform or simply provide new opportunities for extraction.

Western leverage over Ukrainian governance depends on maintaining aid conditionality through the post-war period. This requires political will in Washington, Brussels, and other capitals to sustain pressure even after the existential threat recedes.

Historical experience suggests this is difficult to maintain, post-conflict aid programs often prioritize rapid reconstruction over governance reform, and recipient governments learn to exploit donor fatigue. Ensuring Ukrainian reform will require sustained attention and institutional mechanisms that survive changing political priorities in Western capitals.

DEMOCRACY VERSUS SOVEREIGNTY

The framing "democracy versus authoritarianism" may be less analytically useful than "sovereignty versus imperial subordination."

Ukraine's governance imperfections do not negate Russia's imperial project or the precedent that successful conquest would establish. The fundamental issue is not whether Ukraine meets Western democratic standards but whether nations have the right to exist independent of larger neighbors' claimed spheres of influence.

Russian foreign policy rests on a premise that former Soviet republics lack full sovereignty and remain within Moscow's legitimate zone of control. This view finds expression in official statements, policy documents, and military actions from Georgia in 2008 to Ukraine in 2014 and 2022.

Vladimir Putin's July 2021 essay "On the Historical Unity of Russians and Ukrainians" explicitly denies Ukrainian national identity and statehood, arguing that Ukraine's separation from Russia represents a historical aberration to be corrected.[5]

This imperial ideology poses a direct challenge to the principle of sovereign equality that underpins the modern international order.

If the concern is defending sovereignty rather than democracy per se, Ukraine's governance quality becomes relevant but not determinative. Corrupt, poorly governed states still possess sovereignty rights under international law. Saudi Arabia, Egypt, and numerous other deeply flawed governments enjoy recognized sovereignty without Western attempts at forcible regime change.

The question becomes whether sovereignty rights depend on meeting specific governance standards or exist independent of domestic political arrangements.

This reframing has strategic implications. It suggests that Western support for Ukraine serves the broader interest in maintaining a rules-based order where sovereignty is respected regardless of internal politics. Russian success would establish that powerful states can redefine neighboring countries' sovereignty status based on historical claims, ethnic composition, or geopolitical preference.

Accepting this precedent would destabilize not just Eastern Europe but any region where revisionist powers harbor irredentist ambitions, from the South China Sea to the Persian Gulf to the Horn of Africa.

THE COMPARATIVE AUTHORITARIAN QUESTION

If the concern is avoiding support for corrupt, non-democratic governments, consistency requires examining the full scope of Western partnerships. The United States maintains long-standing alliances with significantly less democratic partners when strategic interests align, Saudi Arabia, Egypt, various Central Asian states, several Southeast Asian countries. These relationships involve military cooperation, intelligence sharing, economic ties, and diplomatic support despite well-documented human rights violations and authoritarian governance.

Ukraine, whatever its flaws, maintains competitive elections (when not prevented by war), opposition parties, civil society organizations, and media outlets that criticize government policy, even if constrained by wartime restrictions.[6]

This places Ukraine closer to democratic governance than many current U.S. partners, suggesting that objections to supporting Ukraine based on governance concerns may reflect inconsistent application of principles.

The inconsistency creates two potential responses.

- One is to reduce support for all authoritarian partners, applying democratic standards uniformly across relationships. This approach has intellectual coherence but would require abandoning partnerships with Egypt, Saudi Arabia, and others, a strategic choice that few policymakers seriously advocate.
- The alternative is to acknowledge that geopolitical interests sometimes require partnerships with imperfect governments while using economic and diplomatic leverage to encourage reform where possible. This reflects traditional realist foreign policy but sits uncomfortably with the moral rhetoric often deployed to justify Western engagement.

The dilemma reveals deeper questions about the relationship between values and interests in foreign policy.

- Can democracies effectively pursue security interests that sometimes require partnering with non-democracies?
- Should strategic calculations override moral principles when they conflict?
- Or can policymakers identify a middle path that advances both security and values through conditional engagement, sustained pressure, and long-term institution building?

These questions have no simple answers, but avoiding them through inconsistent application of democratic standards serves neither strategic clarity nor moral coherence.

THE PRECEDENT DILEMMA

Conditioning support on governance quality establishes a precedent that only perfect democracies merit defense against aggression. This creates perverse incentives for both allies and adversaries.

For adversaries, it suggests that targeting states with governance problems will encounter less Western resistance than attacking consolidated democracies.

For allies, it implies that democratic backsliding may cost them security partnerships, potentially accelerating authoritarian drift as governments conclude they cannot count on external support and must secure their positions through internal repression.

The precedent also raises the question of who determines what constitutes sufficient democratic quality to merit support.

- Does it require fully free and fair elections?
- Independent judiciaries?
- Protected media freedoms?
- Low corruption scores?

If these become formal criteria for alliance membership or security assistance, many current partners would face exclusion, and borderline cases would generate endless diplomatic disputes over whether standards have been met. Creating such criteria also grants adversaries a veto over alliance expansion, undermine a country's governance sufficiently, and it becomes ineligible for Western support.

More fundamentally, linking support to governance quality assumes that corruption or democratic deficits provide legitimate grounds for invasion. If Russia can justify aggression against Ukraine by citing corruption or political problems, what prevents China from making similar arguments about Taiwan, or any powerful state from targeting weaker neighbors whose governance falls short of ideal standards? Or governance standards set by the more powerful power?

The logic that governance quality determines sovereignty rights would fundamentally undermine the principle that borders should not be changed by force, a principle that, however imperfectly observed, has prevented countless wars since 1945.

THE LONG GAME ON CORRUPTION

Corruption in Ukraine reflects structural factors that require generational change rather than immediate solutions. The country inherited Soviet-era governance patterns where formal institutions served as facades for informal networks of patronage and extraction.

Thirty years proved insufficient to dismantle these patterns, particularly given Ukraine's difficult economic transition, oligarchic capture of state institutions, and the ongoing security threat from Russia that diverted resources and attention from reform efforts.

Historical precedents suggest that sovereignty creates the possibility space for reform even when reform proceeds slowly and unevenly. South Korea in the 1960s-1980s combined authoritarian governance, pervasive corruption, and close U.S. alliance relationships. Taiwan followed a similar trajectory. Poland after 1989 inherited deep corruption from the communist period.[7]

All three eventually developed more accountable governance, though the process took decades and involved setbacks.7 Russian control would eliminate this possibility entirely, imperial subjects have far less opportunity to reform their governance than do sovereign states, however imperfect their initial conditions.

The question becomes whether sustained Western engagement can accelerate Ukrainian reform or whether corruption is so deeply embedded that external pressure proves ineffective. EU accession processes provide some evidence for optimism, countries from Romania to Bulgaria to Croatia made significant (if incomplete) governance improvements when seeking membership, responding to concrete conditionality attached to benefits they valued.

Ukraine's EU candidate status creates similar leverage, though whether Brussels will maintain conditionality pressure or accept

Ukraine's membership despite continuing problems remains uncertain.

THE NARRATIVE TRAP

The "democracy versus authoritarianism" framing that dominates Western discourse about Ukraine serves domestic political purposes but creates strategic vulnerabilities. It satisfies the need for moral clarity, simplifies complex situations for public consumption, and mobilizes support by casting the conflict in terms of fundamental values. However, it also sets unrealistic expectations and creates brittleness when Ukraine's imperfections become apparent to Western publics.

When Ukrainian officials engage in corruption, when media freedoms face restrictions, when opposition figures face prosecution, each incident provides fodder for those arguing that Western support lacks moral foundation.

The gap between idealistic rhetoric and messy reality erodes public support more than honest acknowledgment of Ukraine's problems would. Framing support in terms of defending perfect democracy against pure evil creates cognitive dissonance that opponents can exploit, whereas framing it as defending sovereignty against imperial conquest while encouraging gradual reform proves more sustainable when confronted with uncomfortable facts.

A more honest framing would acknowledge Ukraine as an imperfect state facing existential threat from a revisionist power, where Western support serves multiple interests: containing Russian expansionism, maintaining deterrence credibility, defending sovereignty norms, and creating conditions where Ukrainian reform becomes possible.

This rationale lacks the moral simplicity of democracy versus authoritarianism, but it better reflects reality and proves more resilient when confronted with evidence of Ukrainian governance failures.

STRATEGIC PRIORITY
AND MORAL CLARITY

Perhaps the core issue is whether supporting Ukraine serves Western strategic interests regardless of its domestic governance. If the answer is affirmative because Russian defeat shapes adversary calculations globally, strengthens deterrence, depletes Russian military capacity, and reinforces norms against conquest then Ukrainian democracy becomes a secondary consideration. This represents coldly pragmatic realism but may be more intellectually honest than pretending Ukraine is something it currently is not.

The strategic case rests on several foundations.

- First, Russian military failure in Ukraine signals to China, Iran, North Korea, and other revisionist powers that conventional aggression carries high costs and uncertain prospects for success. This deterrent effect operates independently of Ukrainian governance quality.
- Second, the conflict has severely degraded Russian military capabilities through equipment losses, personnel casualties, economic sanctions, and diplomatic isolation thereby reducing Moscow's capacity to threaten NATO members or project power elsewhere.
- Third, Western defense industries have reconstituted production capacity and developed new operational concepts in response to the conflict, improving preparedness for potential future conflicts with peer adversaries.

From this perspective, the billions spent supporting Ukraine represent a strategic investment with returns that extend far beyond Ukrainian borders. The cost of deterring future Russian aggression, containing Chinese ambitions, and maintaining alliance credibility would likely exceed the cost of supporting Ukraine's defense. This calculation holds regardless of whether Ukraine perfectly embodies democratic values—the strategic logic operates separately from the governance question.

However, divorcing strategy from values entirely creates its own problems. If democracies support any government purely on strategic grounds without regard for governance, they abandon the normative framework that distinguishes them from amoral powers pursuing naked interest.

The challenge lies in integrating strategic necessity with values-based foreign policy, acknowledging that interests sometimes require supporting imperfect partners while using that support to encourage improvement rather than simply accepting corruption and authoritarianism as permanent conditions.

CONCLUSION: HOLDING MULTIPLE TRUTHS

The dilemmas surrounding Ukrainian governance and Western support resist clean resolution. They require holding multiple truths simultaneously: that Russian imperialism poses genuine danger and must be defeated; that Ukraine fights for authentic sovereignty, not merely regime survival; that Ukrainian corruption is real, significant, and problematic; and that strategic necessity sometimes means working with imperfect partners while maintaining pressure for improvement.

Neither naive whitewashing of Ukrainian governance problems nor cynical abandonment based on those problems serves Western interests or moral principles. The former creates unrealistic expectations and eventual disillusionment. The latter sacrifices strategic position, alliance credibility, and sovereignty norms to an impossible standard of democratic purity that few current allies could meet.

The sustainable approach acknowledges Ukrainian imperfections while focusing on strategic fundamentals: preventing Russian conquest serves Western interests by maintaining deterrence, containing revisionism, and defending sovereignty norms. This support should include sustained conditionality aimed at encouraging governance reform, particularly in post-war reconstruction.

Western partners should speak honestly about Ukrainian governance challenges rather than pretending they don't exist, while explaining why support remains justified on strategic rather than purely moral grounds.

The test of Ukrainian democracy will come in the post-war period:

- Does the Zelenskyy government relinquish emergency powers?
- Do competitive elections resume with genuine opposition participation?
- Does corruption decrease under pressure from Western partners and Ukrainian civil society?

These questions cannot be answered definitively today, creating irreducible uncertainty about Ukraine's political trajectory.

What can be stated with confidence is that Russian victory would foreclose any possibility of Ukrainian democratic development while establishing dangerous precedents that would destabilize international order.

Western support for Ukrainian sovereignty, distinct from uncritical endorsement of Ukrainian governance, serves both strategic interests and the long-term possibility of reform. This represents not moral compromise but rather strategic maturity: recognizing that security policy must sometimes work with imperfect instruments while maintaining leverage and conditionality to encourage improvement over time.

The alternative, abandoning Ukraine because it fails to meet ideal democratic standards, would not enhance global democracy. It would instead signal that democratic backsliding or governance problems provide justification for conquest, encouraging revisionist powers to target vulnerable neighbors while undermining the sovereignty rights that create the possibility space for reform.

This outcome would serve neither strategic interests nor the values that Western policy claims to defend.

CHAPTER 5
UKRAINE'S
COALITION BUILDING

When Russian tanks rolled across Ukraine's borders on February 24, 2022, few observers predicted that the embattled nation would not only survive but would orchestrate one of the most impressive diplomatic coalitions in modern history.

Over nearly three years of war, Ukraine has built an unprecedented international network spanning more than 50 countries, secured over $145 billion in military assistance, and fundamentally reshaped global perceptions of sovereignty and international law.[1]

This remarkable achievement represents not merely a tactical response to aggression, but a strategic reimagining of how threatened nations can mobilize international support in an era of complex multipolarity.

THE FOUNDATION: BUILDING ON EURO-ATLANTIC RELATIONSHIPS

Ukraine's coalition-building success was not conjured from thin air. The groundwork had been laid over decades of gradual integration with Western institutions. Ukraine joined the North Atlantic Cooperation Council in 1991 and the Partnership for Peace program in 1994, establishing early frameworks for cooperation with NATO.[2]

Following Russia's 2014 annexation of Crimea and intervention in Donbas, these relationships intensified significantly, with NATO establishing trust funds to strengthen Ukraine's defense capabilities, including the Command, Control, Communications and Computers Trust Fund and the Logistics and Standardisation Trust Fund.[3]

The European Union relationship proved equally vital. Ukraine's Association Agreement with the EU, signed in 2014, provided not just economic integration but a political anchor for the country's westward orientation. When the full-scale invasion occurred, this foundation enabled rapid escalation of support. The EU moved with unprecedented speed to grant Ukraine candidate status in June 2022, merely four months after the invasion began, and opened formal accession negotiations.[4]

Yet these pre-existing relationships alone cannot explain Ukraine's diplomatic success. The transformation from regional partner to global cause célèbre required deliberate strategy, exceptional execution, and leadership that understood both the moment's urgency and the game of international coalition management.

THE ZELENSKY FACTOR:
LEADERSHIP IN THE DIGITAL AGE

President Volodymyr Zelensky's decision to remain in Kyiv during the invasion's opening days became a defining moment not just for Ukraine but for the entire conflict's diplomatic trajectory. His now-famous response to American evacuation offers," I need ammunition, not a ride", captured global imagination and established Ukraine's narrative as one of defiance rather than defeat. This single act of courage provided Western leaders political cover to escalate their support and galvanized public opinion across democratic nations.[5]

The Western offer suggested parallels to what happened to the ousted Western governments when the Nazis invaded. Zelensky had a different vision: Ukraine can resist and beat back the Russians? His vision was instrumental in setting the stage for the reversal of Putin's plan for a Crimea II operation.

Zelensky's background as a media professional proved unexpectedly strategic. He understood instinctively how to communicate in the digital age, delivering video addresses to parliaments worldwide, speaking directly to international audiences through social media, and maintaining a constant presence that kept Ukraine's cause at the forefront of global consciousness. This approach represented a new model of wartime diplomacy, transparent, direct, and unfiltered by traditional diplomatic channels.

THE RAMSTEIN FORMAT: INSTITUTIONALIZING SUPPORT

Perhaps Ukraine's most significant institutional achievement has been the creation and maintenance of the Ukraine Defense Contact Group, commonly known as the Ramstein format after its initial meeting location at Ramstein Air Base in Germany. Initiated by then-U.S. Secretary of Defense Lloyd Austin in April 2022, the format brought together 40 nations in its first meeting with a mandate to coordinate military assistance.

The Ramstein format has evolved into far more than a coordinating mechanism. Over 30 meetings (as of October 2025), it has become the primary platform for synchronizing military aid, removing restrictions on weapons transfers, and sustaining political commitment to Ukraine's defense. By September 2025, the coalition had expanded to include 54 countries and committed over $145 billion in military assistance, enabling Ukraine to receive advanced systems including F-16 aircraft, Patriot air defense systems, Leopard and Abrams tanks, and HIMARS artillery systems.[6]

The establishment of specialized coalitions within the broader framework, including tank coalitions, air defense coalitions, and most recently an electronic warfare coalition, allowed countries to contribute according to their specific capabilities while maintaining collective effort.[7]

Critically, leadership of the Ramstein format transitioned in 2025 from the United States to joint UK-German chairmanship, demonstrating the coalition's resilience beyond dependence on any single nation. This transition proved that Ukraine had successfully distributed the diplomatic burden across multiple partners, reducing vulnerability to any one nation's policy changes.[8]

THE PEACE FORMULA: PROACTIVE DIPLOMACY BEYOND DEFENSE

While securing military aid remained paramount, Ukraine recognized early that reactive diplomacy, simply requesting weapons, would be insufficient for long-term coalition maintenance. In November 2022, President Zelensky unveiled Ukraine's 10-point Peace Formula at the G20 summit in Bali, shifting Ukraine from supplicant to agenda-setter in international discussions about the war's resolution.

The Peace Formula addressed issues with global resonance: nuclear safety (particularly regarding the Zaporizhzhia nuclear power plant), food security, energy security, prisoner exchanges, restoration of territorial integrity, withdrawal of Russian forces, prosecution of war crimes, environmental restoration, security guarantees, and confirmation of war's end. This comprehensive approach demonstrated that Ukraine conceived of peace not as mere cessation of hostilities but as addressing the war's systemic impacts on international order.[9]

The diplomatic offensive around the Peace Formula proved remarkably effective at maintaining international engagement. Ukraine organized multiple rounds of consultations with national security advisors from over 80 countries, including meetings in Copenhagen (June 2023), Jeddah (August 2023), and Malta (October 2023).[10] These consultations created working groups for each of the ten points, with specific countries volunteering to champion particular issues, thereby distributing ownership of the peace process across the international community.

The culmination came with the Global Peace Summit in Switzerland in June 2024, which brought together leaders and representatives from nations across the Global North and Global South to discuss implementation of the Peace Formula.[11]

While Russia's absence limited the summit's immediate impact on ending the war, the gathering demonstrated Ukraine's success in positioning itself as the legitimate voice for peace terms, contrasting sharply with Russian proposals that demanded Ukrainian territorial concessions and military limitations.

THE GLOBAL SOUTH CHALLENGE

Ukraine's most difficult diplomatic challenge has been engaging the Global South, nations in Africa, Asia, Latin America, and the Middle East that comprise 40% of UN General Assembly votes and maintain complex relationships with Russia. Many of these countries abstained from UN resolutions condemning Russian aggression, not from sympathy with Moscow but from historical non-alignment traditions, economic dependencies, and resentment of perceived Western double standards.[12]

Ukraine's outreach to these regions has been multifaceted and pragmatic. Foreign Minister Dmytro Kuleba embarked on extensive diplomatic tours, including multiple visits to Africa, sometimes the first such visits by a Ukrainian foreign minister. President Zelensky made unexpected appearances at the Arab League summit in Saudi Arabia in May 2023, directly appealing to Arab leaders and explicitly invoking Ukraine's Muslim community, including Crimean Tatars, to frame the conflict in terms resonant beyond European security concerns.[13]

The "Grain from Ukraine" humanitarian program, launched in autumn 2022, exemplified Ukraine's strategic use of its agricultural resources for diplomatic leverage. By supplying grain to countries particularly affected by the global food crisis, a crisis exacerbated by Russia's blockade of Ukrainian ports and withdrawal from the Black

Sea Grain Initiative, Ukraine positioned itself as a contributor to global food security rather than merely a recipient of aid.

Ukraine also appointed a Special Representative for the Middle East and Africa, signaling institutional commitment to sustained engagement with these regions. Countries including Guatemala, Bahrain, and the United Arab Emirates joined the Crimea Platform, Ukraine's diplomatic initiative focused on the illegal annexation of Crimea, demonstrating gradual if modest progress in building support beyond traditional Western allies.[14]

Yet challenges remain substantial. Russia's narrative positioning itself as an anti-colonial power challenging Western hegemony has found receptive audiences in regions with colonial histories.[15]Russia maintains defense agreements with 30 of 54 African states, provides significant military support to authoritarian regimes, and leverages historical ties from Soviet-era support for liberation movements. Ukraine cannot match Russia's decades of presence in these regions, making quick diplomatic breakthroughs unlikely.

The August 2023 Jeddah meeting on the Peace Formula included participation from Brazil, India, Saudi Arabia, and South Africa, major powers from the Global South, yet their engagement remained measured, with none fully endorsing all aspects of Ukraine's position.

These countries maintain their own strategic interests, including India's energy imports from Russia and China's complex relationship with Moscow. Ukraine's approach has necessarily evolved from seeking full alignment to building issue-specific cooperation where interests overlap, particularly on matters like nuclear safety, food security, and protection of civilians.

THE IMPACT ON THE WAR: MATERIAL AND PSYCHOLOGICAL DIMENSIONS

The coalition Ukraine built has had profound effects on the war's trajectory, both material and psychological. On the material side, the sheer volume of assistance, over $145 billion in military aid alone through September 2025, with NATO allies providing €85 billion in total assistance during 2024-2025, has been decisive in Ukraine's ability to resist a larger, better-equipped adversary.[16]

This support enabled Ukraine to evolve from defensive operations with anti-tank and anti-aircraft missiles in early 2022 to complex combined-arms operations by 2023-2024, incorporating Western armor, artillery, and increasingly sophisticated air defense and precision strike capabilities. The F-16 aircraft, HIMARS systems, and Patriot batteries that seemed politically impossible in March 2022 became reality through persistent diplomatic effort and incremental coalition-building.

Intelligence sharing from Western partners, including satellite imagery, signals intelligence, and real-time battlefield information provided Ukraine asymmetric advantages that partially offset Russia's numerical superiority. Economic assistance prevented Ukraine's economy from collapsing despite massive destruction, maintaining governmental functionality and social services even in wartime conditions.

The psychological impact may be equally significant. The coalition signaled to Ukrainian forces that they were not fighting alone, sustaining morale through the war's darkest moments. For Russian strategic planners, the coalition's durability has complicated calculations about outlasting Ukrainian resistance. The expectation that Western support would fracture or fatigue has repeatedly proven incorrect, forcing Russia to reckon with a Ukrainian state that has staying power backed by substantial international resources.

For the broader international system, Ukraine's coalition-building has established precedents. It demonstrates that aggressive violation of sovereignty and territorial integrity can unite disparate nations in sustained response. It shows that modern diplomacy can mobilize support through digital communication and direct appeals beyond traditional channels. And it proves that even nations under existential threat can exercise diplomatic agency, setting agendas rather than merely reacting to great power decisions.

LIMITATIONS AND
ONGOING CHALLENGES

Despite remarkable achievements, Ukraine's coalition faces significant limitations and ongoing challenges. The coalition has not been universal for large portions of the Global South remain neutral or tacitly pro-Russia. China, though not providing military support to Russia on the scale of Ukraine's Western backing, has maintained strategic partnership with Moscow and declined to condemn the invasion definitively.[17]

Debates over escalation management have created persistent tensions within the coalition. Restrictions on weapons use, particularly regarding strikes on Russian territory with Western-provided systems, have frustrated Ukrainian military planners who argue such limitations hamstring effective defense. Coalition members have moved gradually, often requiring Ukrainian battlefield success to justify the next level of support rather than providing capabilities preemptively.

"War fatigue" represents perhaps the most serious long-term threat to coalition sustainability. As the conflict extends beyond three years, maintaining public support in donor nations has become increasingly challenging, particularly as domestic economic concerns and other international crises compete for attention and resources.

The emergence of different diplomatic approaches, most notably the Trump administration's 2025 initiatives to engage Russia directly in peace negotiations, sometimes without Ukrainian participation, has introduced uncertainty about the coalition's unity going forward.[18]

Financial sustainability poses another challenge. While military assistance has been substantial, Ukraine's reconstruction needs, estimated in the hundreds of billions of dollars, exceed even the generous support provided thus far. Maintaining donor commitment not just for war fighting but for post-war recovery will require sustained diplomatic effort as the conflict eventually transitions to peace.

In short, as the war extends beyond three years, Ukraine faces the challenge of transitioning from crisis-driven coalition maintenance to sustainable long-term partnership architecture. The question is no longer whether Ukraine can build a coalition for it demonstrably has but whether that coalition can endure through the war's potential prolongation and eventual transition to peace.

CHAPTER 6

FROM CINEMATIC
BATTLES TO REAL WAR

In 1970, the rolling hills and vast plains of Ukraine served as the backdrop for one of cinema's most ambitious war epics. Sergei Bondarchuk's "Waterloo" transformed the Soviet republic into 19th-century Belgium, with thousands of Red Army soldiers marching across Ukrainian soil to recreate Napoleon's final defeat.[1]

More than five decades later, those same landscapes echo with the sounds of real warfare, as Ukraine fights for its independence against Russian invasion. The cruel irony of history has turned a cinematic stage into an actual battlefield.

The 1970 film "Waterloo" represented an unprecedented collaboration between East and West during the Cold War era. Italian producer Dino De Laurentiis partnered with Soviet filmmakers to create what would become one of the most authentic war films ever made. The choice of Ukraine as the filming location was both practical and symbolic. Its terrain closely resembled the Belgian countryside where the actual Battle of Waterloo took place on June 18, 1815.

What made the film extraordinary was not just its $25 million budget (enormous for its time), but the Soviet government's unprecedented decision to loan the production 15,000 soldiers from the Red Army. These weren't actors or extras in the traditional sense, but active military personnel who brought genuine martial discipline to their performances. The soldiers were trained for months in 19th-century military tactics, learning to march in formation, handle period weapons, and execute the complex maneuvers that characterized Napoleonic warfare.

The irony was palpable even then: Soviet soldiers, representatives of a communist state that had risen from the ashes of Tsarist Russia, were portraying the armies of Napoleon and Wellington, the very forces that had shaped the Europe their own revolution would later transform. Ukrainian soil, which had witnessed countless real battles throughout history, from Mongol invasions to World War II's devastating Eastern Front, now hosted a carefully choreographed recreation of someone else's war.

Ukraine's selection as the filming location was hardly accidental. The region has long served as a crossroads where empires clash and historical narratives intersect. From the medieval Kyivan Rus to the Polish-Lithuanian Commonwealth, from the Ottoman Empire's northern reaches to the expanding Russian Empire, Ukrainian lands have been contested territory for centuries. The Cossack uprisings, the devastating famines, the Holocaust, and the brutal fighting of World War II, all have left their marks on this soil.

During the Soviet era, Ukraine was both a crucial agricultural heartland and an industrial powerhouse, contributing significantly to the USSR's military and economic might. The soldiers who marched across Ukrainian fields in 1970, recreating Waterloo's charges and counter-charges, were part of this Soviet military machine. Many of these soldiers were themselves Ukrainian, their families having survived Stalin's engineered famines and Hitler's occupation. They were now playing soldiers from a completely different era and conflict, their own complex history temporarily subsumed into the grand narrative of Napoleon's downfall.

The film's production showcased the Soviet Union's organizational capabilities and Ukraine's strategic importance within the federation. The logistics required to coordinate 15,000 soldiers, hundreds of horses, authentic costumes, and period weaponry demonstrated the kind of centralized planning and resource mobilization that characterized Soviet governance. Ukrainian infrastructure, from railways to accommodation facilities, supported this massive undertaking.

"Waterloo" achieved an authenticity that modern CGI-heavy productions struggle to match. The cavalry charges were real, with actual horses and riders thundering across Ukrainian steppes. The artillery smoke that rolled across the battlefield came from genuine cannons firing blank charges. The formations of infantry, thousands strong, moved with the precision that only actual military training could provide.

Bondarchuk, himself a veteran of World War II's Eastern Front, brought a deep understanding of warfare's chaos and brutality to the production. The Ukrainian locations provided the vast open spaces necessary to stage such massive battle scenes. The rolling terrain allowed cameras to capture the ebb and flow of battle across multiple miles, showing how Napoleon's forces gradually succumbed to Wellington's defensive positions and Prussian reinforcements.

The soldiers' performances carried weight because they understood military discipline in ways that civilian actors never could. Their movements during battle scenes reflected genuine tactical knowledge, their responses to commands showed real military training, and their ability to maintain formation under difficult filming conditions demonstrated the kind of unit cohesion that actual armies require.

Yet there was something surreal about watching Soviet soldiers portray the Grande Armée.

Napoleon's multinational force that had included Poles, Germans, Italians, and French. Many of the Ukrainian and Russian soldiers participating in the film came from regions that had actually fought against Napoleon during his 1812 invasion of Russia. They were now, in essence, playing their own historical enemies.

The transformation from cinematic battleground to actual war zone began decades after the cameras stopped rolling. Ukraine's path to independence following the Soviet Union's collapse in 1991 was initially peaceful, but it set the stage for the conflicts that would later engulf the region. The Orange Revolution of 2004, the Euromaidan protests of 2013-2014, and Russia's subsequent annexation of Crimea and support for separatists in eastern Ukraine marked Ukraine's gradual shift from Soviet satellite to contested nation.

The 2022 Russian invasion brought full-scale warfare to Ukrainian soil for the first time since World War II. The same landscapes that had hosted Bondarchuk's elaborate recreations of 19th-century warfare now witnessed 21st-century combat featuring drones, precision missiles, and tank battles that would have been unimaginable to the filmmakers of 1970.

The parallels are haunting.

Where Soviet soldiers once pretended to charge across fields in formation, Ukrainian defenders now fight for their homeland's survival. The Ukrainian steppes that provided such magnificent backdrops for cinematic cavalry charges now see the movement of modern armored columns and the flight patterns of military aircraft.

The coordination and logistics that once served a film production now serve a life-and-death struggle for national existence. The juxtaposition reveals something profound about how we remember and represent conflict.

In 1970, war was something that could be packaged as entertainment, a spectacular historical drama complete with heroic charges and noble defeats. The Battle of Waterloo, safely distant in time, could be transformed into cinematic spectacle. The Soviet soldiers participating were performing someone else's history, playing roles in a conflict that had ended 155 years before they were born.

Today's war in Ukraine carries no such distance or abstraction.

Every explosion, every casualty, every displaced family represents immediate human cost. The Ukrainian soldiers defending their country aren't performing historical roles. They're writing contemporary history with their blood and sacrifice. The soil that once absorbed fake gunpowder and theatrical casualties now bears witness to genuine loss and heroism.

The film "Waterloo" captured Napoleon's famous observation

that "from the sublime to the ridiculous is but a step." The current reality in Ukraine suggests a different truth: from the cinematic to the real is also but a step, and that step can be measured in human lives rather than entertainment value.

The 15,000 Soviet soldiers who participated in filming "Waterloo" have long since returned to civilian life or passed away. Many were Ukrainian, and some of their descendants may now be fighting in the current conflict. The transformation from extras in a historical drama to participants in contemporary history represents one of those strange turns that only time can create.

The film itself remains a masterpiece of historical recreation, its battle scenes unmatched in their scope and authenticity. Yet viewing it today, knowing what has transpired in Ukraine, adds layers of meaning that its creators never intended. Every Ukrainian face among the Soviet extras, every piece of Ukrainian terrain captured on film, now carries the weight of current events.

Perhaps this is what makes the comparison so compelling: it reminds us that history is never truly past, that the landscapes we use to remember old wars can quickly become the sites of new ones.

The Ukrainian fields that once echoed with the choreographed sounds of Napoleonic battle now ring with the urgent reality of contemporary conflict. The transition from performance to reality, from recreation to creation of new history, serves as a sobering reminder that peace, like the elaborate staging of old wars, can be more fragile than we imagine.

The story of Ukraine, from Soviet film set to independent nation under siege, encapsulates the unpredictable nature of history itself. What begins as entertainment can become earnest reality; what seems like distant past can suddenly become urgent present.

In this transformation lies both tragedy and testament to the enduring human capacity for both creation and destruction, for both artistic achievement and actual heroism.

SECTION IV
THE EUROPEAN DYNAMIC

CHAPTER 1
EUROPE'S DEFENSE RENAISSANCE

Europe stands at an inflection point. After decades of strategic drift, the continent faces a fundamental choice: develop genuine strategic autonomy and rebuild credible military capabilities, or accept permanent vulnerability in an increasingly hostile multipolar authoritarian world.

Russia's full-scale invasion of Ukraine in 2022 didn't create this challenge. It merely stripped away the comfortable illusions that had sustained European security policy since the Cold War's end.

This transformation extends far beyond defense budgets and force structures. It requires reimagining strategic culture across the continent, embedding defense within the fabric of European societies, and forging new models of international cooperation that move at the speed of modern warfare rather than traditional procurement cycles.

My late colleague and friend, Brendan Sargeant, highlighted the importance of reworking strategic culture in times of profound historical change. He put it this way. "In times of great change, the challenge is to imagination, for continuity in strategy is likely to lead to failure."[1]

This is the challenge which faces Europe. The stakes are existential: Europe's choices will determine whether democratic values can survive the coordinated challenge from authoritarian powers, or whether the continent becomes a strategic object rather than a strategic actor.

HOW EUROPE FORGOT
HOW TO DEFEND ITSELF

Germany's trajectory illustrates Europe's broader strategic journey. During the 1980s, West Germany maintained credible defense capabilities built on three pillars: NATO integration, close U.S. partnership, and commitment to "peace through strength." The Air-Land Battle doctrine exemplified this approach, using superior coordination, technology, and tactics to offset numerical disadvantages against Warsaw Pact forces. This strategy emphasized deep battle concepts: engaging enemy front lines while simultaneously striking reserves, supply lines, and command centers.

Crucially, Germans understood that reunification required credible defense capabilities, strong alliances, and persistent pressure on Soviet leadership. When the Berlin Wall fell in 1989, it happened because of military strength backed by diplomatic persistence, not wishful thinking about inevitable progress. The two-track approach to nuclear modernization, which led to the INF treaty with Soviet leadership, demonstrated how military capability and diplomatic engagement could work in concert.

I spent much of time in the 1980s in Germany working on these issues and dealing with European governments on the Euromissile and related issues. And I saw first hand, the level of commitment of West Germans to the defense of their part of Germany against a formidable threat.

I chaired a working group at the Institute for Defense Analysis from 1986 on about how to imagine a process to deliver German reunification. This was a living example of Sargeant said is necessary in periods of profound change. And we are certainly there again.

Victory in the Cold War, however, paradoxically undermined

European defense thinking. The Clinton Administration focused on expanding NATO and the European Union while integrating China into the global economy, decisions that planted seeds for today's multipolar authoritarian challenge. The Clinton and Obama Administrations had no real focus upon the direct defense of Europe and worked hard to shift European militaries to out of area operations.

Germany embraced its role as Europe's dominant economic power while treating defense spending as an unnecessary cost. The assumption that "friends surrounded us" became conventional wisdom, encoded in Germany's 2006 White Book.

Chancellor Angela Merkel epitomized this approach from 2005 to 2021. Her strategy centered on German leadership of EU economic decisions while maintaining minimal defense commitments. As a former East German who spoke Russian, she believed Germany could work with Putin as a rational actor within the globalization framework. Defense was dramatically marginalized in shaping Germany's geopolitical strategy.

This wasn't unusual. During the 2010s, the West worked with Putin on various deals assuming Russia was willing to be part of a broader globalization strategy. The most remarkable was France's agreement to sell Russia two Mistral-class amphibious assault ships, a €1.2 billion contract signed in 2011 that included full technological transfer. Both Sarkozy and Merkel missed the point that Putin was actually a new Tsar, not a cog in the globalization machine.

Meanwhile, Washington was preoccupied with Middle East land wars, encouraging NATO members to contribute to "out of area" missions because European defense was supposedly no longer central. Illustrative of the Obama Administration's priorities was its 2011 decision to dissolve the fleet serving North Atlantic missions, only restored in 2018 under President Trump.

CRIMEA AS STRATEGIC WAKE-UP CALL

Putin's 2014 seizure of Crimea marked the beginning of Europe's strategic awakening, though the full implications took years to register. As Brigadier General (ret.) Rainer Meyer zum Felde characterized it in an interview with me:

> *After unification, the basic belief was that friends surrounded us. We wrote in the 2006 White Book that we did not face a direct threat from Russia anymore. But the Russian aggression against Ukraine made clear that this was a wrong assessment. Russia is back as a threat to Europe in the short and midterm. And what we have to be concerned about in the long run is an emerging axis between Russia and China, ganging up on a global scale against the West.*[2]

The challenge for Germany and much of Western Europe was rebuilding high-end military capabilities in a social context where many citizens did not share NATO governments' assessment that Russia posed a direct threat. This disconnect between strategic reality and public perception would persist until February 2022, when Russia's full-scale invasion finally shattered remaining illusions.

UKRAINE AS MILITARY INNOVATION LABORATORY

The 2022 invasion forced Europe into uncharted strategic territory. Among the most significant developments has been the drone revolution, with Germany and the UK emerging as leaders in a new model of defense cooperation that prioritizes rapid innovation over traditional procurement cycles.

Germany's collaboration with Ukraine on drone warfare represents the most significant wartime defense cooperation in modern history. German companies have delivered over 900 advanced drones while establishing complete manufacturing facilities on Ukrainian soil. Key contributions include Quantum-Systems' nearly 500 Vector reconnaissance drones with AI-powered target detection, Helsing's commitment to 10,000 AI-enabled strike drones, Rheinmetall's SurveilSPIRE reconnaissance systems and Skynex air defense platforms, and HENSOLDT's 10 TRML-4D radars capable of tracking 1,500 targets simultaneously up to 250 kilometers range.[3]

This partnership has fundamentally altered German military thinking. Defense officials now recognize that drones are a key element of modern warfare, validating the shift toward "precision mass" where large quantities of AI-enabled systems can be leveraged rather than relying simply on a small numbers of exquisite platforms. Combat data is striking: German-supported systems maintain 67% survival rates in intensive electronic warfare environments while achieving mission objectives.[4]

The UK has pursued a parallel path through "Project Octopus," announced at the DSEI 2025 defense exhibition in London. This landmark partnership commits Britain to mass-producing thousands of Ukrainian-designed interceptor drones monthly, a fundamental shift from traditional arms transfers to manufacturing Ukrainian-designed technology while sharing intellectual property and industrial expertise.[5]

The economic calculus driving this transformation is compelling. Each Russian Shahed drone costs approximately $35,000-$50,000, while traditional interceptor missiles like the Stinger cost $120,000-$150,000, and advanced Patriot PAC-3 interceptors cost about $4 million per unit. In contrast, Ukrainian companies are producing Shahed interceptors at roughly $5,000 each, with some developmental models costing as little as $300.[6]

The battlefield effectiveness validates this approach. On July 23, 2025, Ukrainian media reported that 9 out of 10 Shahed drones shot down were due to interceptor drones rather than traditional air defense systems. President Volodymyr Zelensky confirmed that four Ukrainian companies manufacture interceptor drones, with two already achieving significant battlefield success.[7]

Britain's commitment represents unprecedented production scale. The UK will invest £350 million in 2025 to increase drone supplies to Ukraine from 10,000 in 2024 to 100,000 in 2025, a ten-fold increase. This includes significant industrial investment, with Ukrainian drone manufacturer Ukrspecsystems investing £200 million in a new 11,000-square-meter factory in Mildenhall, Suffolk, expected to create up to 500 jobs with production beginning in early 2026.[8]

Germany's support includes €5 billion in defense cooperation agreements and €400 million investment in Ukrainian long-range drone production. These represent more than commercial relationships. They're strategic industrial partnerships that enhance both nations' capabilities. Germany has pioneered joint production models and evolved from restrictive to enabling export policies for drone technologies, now supporting the "Danish Model" of direct procurement from Ukrainian defense industry.[9]

The transformation extends to dual-use technology integration. German companies are incorporating civilian AI advances, commercial manufacturing techniques, and startup innovation models into defense applications, a significant departure from traditional segregation between commercial and military sectors.

FIFTH-GENERATION AIR POWER: GERMANY'S ENTRY INTO THE F-35 ENTERPRISE

Germany's 2022 decision to acquire 35 F-35 fighters represents more than aircraft procurement. The F-35 global coalition enables significant fighting capability as an integrated force, with Israeli operations providing proof of concept beyond any doubt. For Germany, the transition involves two critical elements: transforming the German Air Force to leverage this new capability for the defense of Germany itself, and becoming a plank holder in an entirely new European coalition capability.

Germany joins a network that includes Italy, which has emerged as Europe's leader in this domain. Italy now operates Europe's best air force, with ability to build more combat aircraft than any other European country and having built the leading air training center in Europe in Sardinia.

Germany actively participates in the Italian International Flight

Training School (IFTS), sending Luftwaffe pilot candidates to undergo advanced training phases at the Italian Air Force's facilities. This collaboration, formally established through an agreement between the chiefs of the German and Italian air forces, allows German pilots to undertake the Phase IV "Lead-In to Fighter Training" at the 61st Wing Airbase in Lecce-Galatina, with operations now expanding to the state-of-the-art IFTS campus at Decimomannu Air Force Base in Sardinia.

The training leverages the T-346A (M-346) advanced jet trainer and integrates modern simulation technologies, providing German trainees with immersive instruction in both synthetic and in-flight modules. This partnership is a recognition of the Italian system's excellence and evidences Germany's intent to potentially deepen involvement to earlier training phases in the future, underscoring the IFTS's stature as a hub for high-level multinational military aviation education.[10]

BUILDING THE NEW SECURITY ARCHITECTURE: SKY SHIELD AND THE DRONE WALL

In response to Russia's invasion and escalating security threats, Germany has emerged as a leading force in reshaping Europe's defense landscape through two ambitious initiatives that represent fundamental shifts in European defense thinking.

The European Sky Shield Initiative (ESSI), originally proposed by German Chancellor Olaf Scholz in August 2022, stands as one of the most significant multilateral defense projects in recent European history. Designed as a ground-based integrated European air defense system with anti-ballistic missile capabilities, the initiative aims to bolster European air and missile defense through joint acquisition by European nations. As of 2025, 24 European states participate, making it one of NATO's most comprehensive collective defense projects.[11]

The system's technical architecture relies on a multi-layered approach: Israel Aerospace Industries' Arrow-3 system for long-range interception, U.S. Raytheon Patriot missiles for medium-range threats, and the German-made Diehl Iris-T system for short-range targets. This configuration reflects both the urgent need for imme-diate capabilities and current limitations of purely European defense technologies. NATO Deputy Secretary General Mircea Geoană emphasized that the initiative "shows the value of Allies stepping up to meet NATO's requirements, while ensuring interoperability and integration."[12]

Parallel to ESSI, Germany is spearheading NATO's "Drone Wall", a continuous network of unmanned aerial systems stretching from Norway to Poland, covering approximately 1,850 miles of NATO's most vulnerable frontier. Announced in 2024, the Drone Wall responds to Russia's use of unconventional warfare tactics, including drone incursions, GPS jamming, and cross-border provocations. Lithuania's Interior Minister Agnė Bilotaitė described it as "a completely new thing — a drone wall stretching from Norway to Poland, and the goal is to use drones and other technologies to protect our borders."[13]

Built around a layered system of AI-powered reconnaissance drones, ground-based sensors, mobile counter-drone platforms, and satellite surveillance, the system aims to detect and disrupt threats in real-time while providing NATO with faster, more accurate intelli-gence across its most exposed borders. Germany leads the project with backing from six NATO countries: Estonia, Latvia, Lithuania, Finland, Poland, and Norway.

The strategic rationale extends beyond mere technological capability. An air defense shield would prevent Russian missiles and drones from striking NATO territory, potentially triggering Article 5's mutual defense clause. In this sense, the shield serves a de-escalatory function by averting crises that could spiral out of control.

The Drone Wall specifically addresses gray-zone warfare, where traditional deterrence mechanisms prove insufficient. As Gundbert Scherf, CEO of German AI defense firm Helsing, explained: "If we deploy there in large numbers, rely on asymmetric capabilities and concentrate tens of thousands of combat drones there, then it will be a very credible conventional deterrent."[14]

THE ANGLO-FRENCH RENAISSANCE: REVIVING THE ENTENTE CORDIALE

While Germany shapes continental European defense architecture, Britain and France have moved to revive their historic partnership in response to the changed security environment. President Emmanuel Macron's July 2025 state visit to London marked a turning point in cross-Channel defense relations, weakened by Brexit but now urgently needed in face of Russian aggression and uncertainty over U.S. commitment to NATO.[15]

The visit resulted in the Northwood Declaration, seeking for the first time to coordinate Britain and France's independent nuclear deterrents. "From today, our adversaries will know that any extreme threat to this continent would prompt a response from our two nations," Prime Minister Keir Starmer declared. The summit created an Anglo-French "oversight committee" to coordinate nuclear capabilities, stronger cooperation in response to Moscow's use of nuclear threats after its 2022 invasion.[16]

French air force general Bruno Cunat emphasized that Britain and France, as European nuclear powers, have a responsibility to demonstrate "strategic solidarity" for European security and "modernize" their cooperation. The Russian nuclear threats made it necessary for the two nations to manage Moscow's "vertical escalation" of nuclear threat.[17]

Beyond nuclear cooperation, the summit addressed modernization of the combined joint force, renamed from the Combined Joint European Force (CJEF). Originally planned as an expeditionary force, the threat is now seen as high intensity war on the European continent. The modernization seeks to make the Anglo-French force more robust, faster to respond, and open to friendly forces, 10 times larger at 50,000 troops strong and capable of responding in every domain.

German Chancellor Friedrich Merz followed in Macron's steps with a one-day official visit to the UK, signing the Kensington treaty, a "friendship treaty" seeking to deepen military and bilateral economic links. The unprecedented almost-alliance between Berlin and London was prompted by Russian invasion of Ukraine.[18]

The Anglo-German pact includes a push for export sales of Eurofighter Typhoon jets and Boxer armored vehicles, and discussion of developing a deep-strike missile with range exceeding 2,000 km, a European long-range capability following Ukrainian use of U.S.-built HIMARS systems.[19]

FRANCE'S STRATEGIC AWAKENING: FROM SOFT POWER TO HARD DETERRENCE

France's transformation under President Macron illustrates how deeply the security environment has shifted European strategic thinking. On the eve of Bastille Day 2025, Macron announced adding €3.5 billion to the 2026 defense budget and €3 billion in 2027, hitting in 2027 the spending target previously set for 2030. The total budget of €64 billion, double the €32 billion defense budget for 2017, represents a dramatic acceleration of French military investment.[20]

Macron's speech emphasized French production of weapons offering "European solutions" for rearming Europe: the new generation SAMP/T missile system, wide range of missiles, Rafale fighters, the first constellation of low-orbit satellites, artificial intelligence, radar, and anti-drone systems. "Let us buy European in volume," he declared, adding that European power was the best shield in face of uncertainty over the U.S., China, and the Russian threat.[21]

Macron's keynote speech crystallized this new French strategic culture: "In the end, let us be clear: to be free in the world, one must be feared. To be feared, one must be powerful." This represents a fundamental departure from the soft power approach that characterized much of post-Cold War European thinking.

THE STRATEGIC IMPERATIVE: EUROPE'S CHOICE

Europe's defense renaissance represents the continent's best opportunity to achieve genuine strategic autonomy while maintaining Atlantic partnership. The combination of fifth-generation air power, AI-enabled drone capabilities, integrated air defense systems, and revitalized industrial partnerships provides the technological foundation for credible deterrence.

Success requires overcoming traditional reluctance to lead militarily and embed defense thinking within broader strategic culture and not treat it as diplomatic afterthought. This means sustained investment, public education, and political commitment across multiple election cycles. Germany's transformation is particularly crucial: as Europe's largest economy and most populous nation, German choices will shape whether Europe can defend itself or remains dependent on external powers for security.

The authoritarian powers are watching carefully. Russia's imperial ambitions and China's global expansion will test European resolve repeatedly. The question is whether Europe will seize this moment with the same strategic imagination that built the Atlantic Alliance and secured German reunification, or drift back into comfortable assumptions that proved illusory.

Several factors suggest this transformation may prove more durable than previous European defense initiatives.

- First, the threat is visceral and immediate. Russian forces are actively engaged in the largest land war in Europe since 1945, with Ukraine serving as buffer and battlefield.
- Second, the innovation emerging from Ukraine provides battle-tested technologies and operational concepts that work, not theoretical capabilities.
- Third, new partnership models like UK-Ukraine and Germany-Ukraine drone cooperation demonstrate that

rapid innovation cycles are possible when traditional
procurement bureaucracy is bypassed.
- Fourth, the demographic and economic pressures on
  Europe make strategic autonomy increasingly necessary.
  The United States faces its own strategic challenges in the
  Indo-Pacific and growing fiscal constraints. European
  nations can no longer assume automatic American
  security guarantees or that Washington will prioritize
  European security over its own interests. This reality
  creates both urgency and opportunity for genuine
  European strategic autonomy.

The infrastructure investments such as the European Sky Shield
Initiative, the NATO Drone Wall, new industrial facilities, and
modernized command and control systems represent physical
commitments that will shape European security for decades. Unlike
budget increases that can be reversed in future political cycles, these
concrete capabilities establish new baseline expectations for what
European security requires.

Yet significant obstacles remain.

- European defense industries still lack the scale and
  integration of American counterparts.
- Fragmented national procurement processes create
  inefficiencies and redundancies.
- Political will remains uneven across member states, with
  some nations more threatened by Russia than others and
  therefore more willing to prioritize defense spending.
- The challenge of coordinating 27 EU member states plus
  additional NATO partners creates complexity that
  adversaries can exploit.

Perhaps most fundamentally, European societies must overcome
decades of strategic drift and reconnect defense with broader questions of sovereignty, values, and collective identity. The comfortable

assumption that military power was obsolete in the modern world has proven catastrophically wrong.

Rebuilding societal consensus around the necessity of credible military capabilities will require political leadership willing to make sustained arguments about security threats and defense requirements.

And the challenge to development of an appropriate strategic culture to deal with the challenge from the multi-polar authoritarian world remains.

The kind of strategic imagination envisaged by Brendan Sargeant is a key part of shaping a European approach which recognizes that security and defense are not externalities to be paid outside of "normal" life. Rather, they are an integral part of shaping a viable economy, society and polity able to be resilient in the face of 21st century authoritarianism.

CHAPTER 2

INFRASTRUCTURE AS BATTLESPACE: THE RETURN OF DIRECT DEFENSE IN EUROPE

In 2020, Murielle Delaporte and I published a book examining what we termed "the return of direct defense" in Europe. Our central argument challenged conventional thinking about European security: defending Europe in the 2020s requires moving beyond traditional military deterrence to embrace a broader strategic concept that places infrastructure, resilience, and supply chains at the center of the contest with 21st-century authoritarian powers. Five years later, events have validated this analysis in ways we could not have fully anticipated.

The transformation we described represents a fundamental shift in how democracies must think about defense. Infrastructure, ports, energy networks, data cables, telecommunications systems, transport corridors, and digital platforms, is no longer merely the backdrop to military operations. It has become both target and weapon in a long-term strategic competition between liberal democracies and authoritarian states that understand power operates across multiple dimensions simultaneously.

FROM COLD WAR TEMPLATES
TO 21ST CENTURY THREATS

The late Cold War model of European defense was architecturally elegant in its clarity. NATO planned around a central front in Germany, clearly defined flanks, and the integration of massive nuclear and conventional forces to deter a Soviet armored thrust westward. The infrastructure of defense, bases, depots, communications networks, mattered enormously, but primarily as the physical substrate supporting armies, air forces, and nuclear arsenals. The threat was identifiable, the geography was stable, and the Alliance's mission was unambiguous.

Today's challenge bears little resemblance to that framework. Russia's 2014 seizure of Crimea and the parallel emergence of China as a system-shaping power revealed an authoritarian toolkit that operates across a far broader spectrum. Contemporary Russia exploits the newly independent states around its "window to the West" not through massed armor columns but through cyber operations, information warfare, energy dependency, and the strategic ownership of critical assets. These instruments allow Moscow to influence European decision-making and constrain policy options well short of open warfare.

China's approach differs in execution but shares this fundamental logic: economic penetration, gray-zone activities, and targeted control of infrastructure reshape the external environment without conventional military campaigns. Both powers have learned to weaponize interdependence, turning the openness of liberal market economies into strategic vulnerabilities that can be exploited for political effect.

THE INFRASTRUCTURE
CHALLENGE: CYBER PENETRATION
TO LEVERAGED ACQUISITION

When we examined the "infrastructure challenge" in our book, we argued that it represents one of the central strategic shifts reshaping contemporary warfare and deterrence, not an afterthought to "real" military planning. The authoritarian toolkit extends from sophisticated cyber penetration of networks to what we termed "leveraged acquisition": the systematic purchase or control of critical assets and connectivity nodes inside liberal democracies.

Chinese and Russian actors, often operating through ostensibly commercial entities, acquire stakes in ports, logistics hubs, energy companies, telecommunications infrastructure, data centers, and high-end manufacturing facilities. This creates enduring political and security leverage from within target societies. The threat is not simply espionage or intellectual property theft; it is the capability to threaten disruption or denial of crucial services during a crisis, to shape elite decision-making through economic dependency, and to gain operational insight into European patterns and vulnerabilities.

This connects directly to the broader concept of hybrid and gray-zone operations. In this operational model, lethal military force often serves as a supporting element while influence over infrastructure, information systems, and supply chains constitutes the main effort. Infrastructure is no longer neutral terrain: it has become a principal battlespace in great power competition.

The security implications cascade across multiple domains. A telecommunications network owned or substantially controlled by Chinese entities creates surveillance opportunities, data integrity risks, and potential denial of service during a crisis. A port facility in which Russian-linked companies hold significant stakes offers insight into military and commercial logistics flows while providing leverage over regional transport networks. Energy infrastructure that creates dependencies on authoritarian suppliers becomes a political weapon, as Europe discovered painfully after Russia's full-scale invasion of Ukraine in 2022.

NATO, THE EU, AND THE DUAL ALLIANCE PROBLEM

The return of direct defense creates what we characterized as a "dual alliance problem." NATO, with its focus on military capabilities, exercises, and collective defense commitments, cannot alone address vulnerabilities in energy grids, data networks, transport corridors, or medical supply chains. The European Union, historically weak on traditional defense matters, holds crucial instruments for infrastructure regulation, economic security, and coordinated crisis management. Neither institution can solve the problem in isolation.

The 2014 shocks, Russia's seizure of Crimea and the mass migration flows from the Middle East and North Africa, exposed fundamental European weaknesses in controlling borders, managing crises, and protecting critical systems. The COVID-19 pandemic later revealed the extent of European dependence on Chinese-dominated supply chains for essential goods, particularly pharmaceuticals and medical equipment. These successive crises underscored the need for what we and other strategic thinkers have termed "smart sovereignty": the capability to guarantee critical supplies and shape diversified, trusted production networks rather than simply purchasing at the lowest cost on global markets.

This represents a profound conceptual shift. Sovereignty in the 21st century is not merely about territorial control or diplomatic recognition. It includes the ability to maintain societal functions and military operations without catastrophic dependencies on potential adversaries. This insight has direct implications for how democracies structure their economies, screen foreign investments, and organize crisis management.

SUPPLY CHAINS AS STRATEGIC WEAPONS

Our research, including extensive interviews with Nordic and European defense experts, drove home a central point: supply chain security and resilience planning, once treated as purely economic or civil protection concerns, must now be understood as core defense functions. The assumptions underlying just-in-time logistics and globally optimized supply chains collapse under crisis conditions and would fragment even more severely during serious geopolitical conflict.

The evidence is stark. When crisis strikes, nations prioritize their own populations. Supply chains fracture at unpredictable points, from obscure component suppliers to shipping chokepoints. The complexity of modern manufacturing means that a single specialized facility in one country can become a strategic bottleneck affecting military production and essential civilian goods across an entire continent.

The strategic response must operate on multiple levels.

- First, predictable essentials must be stockpiled, following models developed by Finland and historically practiced by Sweden and Denmark.
- Second, alternative production capacity, including advanced manufacturing techniques like 3D printing and rapid conversion of existing production lines, must be developed and regularly tested. The crucial caveat: high-end systems such as precision-guided weapons and advanced medical devices cannot be improvised under crisis conditions. They require sustained investment in production capacity and supply chain security during peacetime.
- Third, serious cross-national planning must map likely supply chain vulnerabilities and identify alternative sourcing and production paths before crises hit. This cannot be purely national work; smaller European states in particular must collaborate in regional clusters and through EU frameworks to build credible supply chain security and crisis management mechanisms capable of functioning under stress.

NORDIC MODELS: RESILIENCE AND SMART SOVEREIGNTY

Finland emerged in our research as a compelling model for "smart sovereignty" in an age of hybrid threats. Finnish national security planning emphasizes comprehensive defense: stockpiles, conscription, territorial defense, and deep civil-military integration. Finnish officials described a holistic security-of-supply system covering not only ammunition and fuel but also electricity, telecommunications, data integrity, and physical shelters for both civilians and critical infrastructure.

The underlying logic is both simple and profound: in a compressed crisis, states must fight and function with what they already possess. Reliance on overseas replenishment becomes unrealistic when adversaries actively create anti-access and area-denial conditions across physical, cyber, and financial domains. Data security, ensuring both availability and integrity of essential information systems, now constitutes a distinct layer of infrastructure defense, with dedicated agencies tasked to protect national datasets from corruption or manipulation.

This concept of resilience is not merely supplementary to traditional deterrence: it is a central component. Authoritarian adversaries specifically target weak points in infrastructure and civil society precisely because these are the levers most likely to generate political effects without triggering NATO's Article 5 collective defense commitments. A society that can absorb shocks, maintain essential functions, and continue operations despite disruption is significantly harder to coerce than one that depends on fragile, easily interdicted systems.

The institutional response across Europe has included important innovations. The European Centre of Excellence for Countering Hybrid Threats in Helsinki represents one attempt to pool knowledge and help EU and NATO members develop capabilities for responding to influence operations, infrastructure attacks, and cross-domain pressures that blend economic, informational, and political instruments below the threshold of conventional warfare.

Our research suggested that effective infrastructure defense will not emerge from purely national efforts or exclusively supranational approaches. Instead, it will grow from coalitions of states with shared threat perceptions, the "clusters" we identified, operating under EU and NATO frameworks but not waiting for full alliance-wide consensus before taking action. This logic applies across domains: protection of undersea cables, screening of foreign investments in critical sectors, coordinated responses to disinformation campaigns, and joint development of resilience capabilities.

THE MARITIME
INFRASTRUCTURE FRONT

The concept of a "fourth battle of the Atlantic" illustrates how infrastructure defense, geography, and high-end warfighting intersect in practice. Russia's military modernization around the Kola Peninsula, its advanced submarine and long-range missile forces, and its focus on denying NATO access across the North Atlantic directly threaten European ports, logistics hubs, and undersea communication and energy infrastructure.

Allied responses, new maritime patrol aircraft, F-35-enabled networked air operations, hardened and dispersed air bases, revitalized anti-submarine warfare capabilities, represent not merely force modernization but measures to protect the infrastructural lifelines connecting North America and Europe. Modern maritime contestation now includes hybrid elements: potential attacks on submarine cables, GPS jamming, cyber operations against shipping and port management systems, and the use of civilian vessels for military purposes.

In this environment, infrastructure protection, harbor facilities, fuel depots, data cables, and automatic identification systems, becomes integral to deterrence. Losing control of these nodes rapidly undermines the ability to surge forces or sustain operations. The

Atlantic is simultaneously a military theater and a complex infrastructure system that must be defended as an integrated whole.

FROM CRISIS MANAGEMENT TO OPERATIONAL CAPABILITY

The COVID-19 pandemic provided a brutal stress test for European infrastructure and crisis governance. Both the EU and national governments had assumed global markets would provide essential goods under all conditions. Instead, supply chains broke down completely and voluntary "solidarity" between member states proved insufficient when all faced simultaneous crises. Proposals for EU-level strategic stockpiles and regional hubs of critical equipment were important, but we argued they must be anchored in serious national planning and cross-border coordination rather than treated as substitutes for it.

This connects to a broader legitimacy question for the European Union. If Brussels cannot demonstrate real value in managing transnational crises, whether pandemics, migration, or infrastructure shocks, its credibility as a security actor will erode. Infrastructure defense and supply chain security thus become litmus tests of whether the EU can move beyond regulatory authority to operational relevance, particularly when authoritarian powers actively exploit economic dependencies and political divisions across the Union.

The European Defence Fund and related initiatives represent attempts to reduce fragmentation in European defense industries and stimulate collaborative development of key capabilities, including those relevant to infrastructure protection, cyber defense, and secure communications. For such programs to contribute meaningfully to direct defense, they must target genuine capability gaps identified in both EU and NATO planning processes rather than simply subsidizing national industrial champions.

Investment screening mechanisms have emerged as another critical tool. However, Europe's economic difficulties and internal political divisions create constant pressure to welcome external capital regardless of strategic implications. Credible infrastructure defense requires not only legal frameworks for screening foreign investments but also sustained political will to prioritize long-term security over short-term financial attraction.

THE POLITICS OF INFRASTRUCTURE SECURITY

A central political theme in our analysis was that Europe's response to authoritarian pressure will be shaped less by centralized grand designs than by the interaction of diverse national and regional trajectories. We identified multiple dynamics: Brussels-driven efforts toward convergence and common standards; national reassertion driven by migration, economic stress, and cultural politics; and a "clusterization" trend in which groups of like-minded states form practical coalitions on specific security issues.

Infrastructure defense sits at the nexus of these tensions. Decisions about ports, telecommunications networks, energy grids, and data centers embody fundamental questions about the relationship between national sovereignty, EU-level regulation, and regional solidarity.

The Nordic countries offer one model: shared work on security of supply, cross-border air and maritime exercises, common technical standards, and political alignment in confronting Russian pressure. Similar patterns could emerge elsewhere in Europe, creating overlapping networks of trusted partners capable of rapid action to protect infrastructure and supply chains even when broader alliance consensus proves elusive.

CONCLUSION: DEFENSE TRANSFORMED

Our central conclusion in 2020 was that direct defense has returned to Europe, but in a form far more complex than Cold War precedents. The authoritarian challenge extends well beyond armored offensives or missile salvos to include sustained efforts to penetrate and shape

European infrastructure, information environments, and supply chains. NATO's military adaptation, the EU's emerging role in crisis management and industrial policy, and the growing emphasis on national and regional resilience all converge on a single strategic imperative: defend the systems that allow European societies and armed forces to function.

This transformation requires Europeans to fundamentally rethink sovereignty, alliance roles, and the balance between economic openness and security. Infrastructure defense, understood broadly to encompass energy, data, transport, medical and industrial supply chains, and societal resilience, is not a technical or secondary concern. It has become a central battlefield in the long contest between liberal democracies and 21st-century authoritarian powers.

The challenge is formidable, but it is also clarifying. When infrastructure becomes battlespace, defense becomes everyone's responsibility. Military forces, intelligence agencies, regulatory bodies, private companies, and civil society organizations all have roles to play. Success will require unprecedented coordination across traditional boundaries while maintaining the adaptability to respond to threats that evolve faster than bureaucratic consensus.

The authoritarian powers have already adapted their strategies to exploit the infrastructure dependencies of open societies. The question is whether democracies can respond with sufficient speed, coherence, and political will to defend the systems on which their security and prosperity ultimately depend.

CHAPTER 3
THE NORTHERN SHIELD

When 1,500 Lithuanian paramilitaries descended on the industrial city of Utena in October 2025, staging large-scale urban warfare drills complete with simulated explosions and checkpoint operations, they demonstrated something profound: the nations closest to Russia have fundamentally transformed how democracies prepare for potential conflict.[1]

While Western European allies debate defense spending percentages, the Baltic states, Nordic countries, and Poland have moved beyond theoretical commitments to operational readiness.

This transformation represents more than military buildup. These nations are pioneering a whole-of-society approach to defense that integrates civilian resilience, total defense concepts, hybrid warfare countermeasures, and conventional military capability in ways that could reshape NATO's entire strategic posture.

Their innovations emerge from geographic proximity to threat, historical memory of occupation, and acute awareness that they would form the first line of defense in any confrontation with Russia.

THE BALTIC LABORATORY: FROM VULNERABILITY TO RESILIENCE

The three Baltic states, Estonia, Latvia, and Lithuania, share 750 miles of border with Russia and Belarus. Their combined population barely exceeds six million. Geography and demography make them acutely vulnerable, yet they've transformed potential weakness into strategic innovation.

Lithuania's Rifleman's Union exemplifies this approach. Founded during the nation's early 20th-century independence struggle, the organization has expanded to 17,000 members, nearly matching the 20,000-strong regular Lithuanian Armed Forces. Members are ordinary citizens: parents, professionals, teachers who dedicate evenings and weekends to military training.

This isn't symbolic patriotism. The LRU trains to defend critical infrastructure, establish checkpoints, impose curfews, and conduct urban defense operations. During the Utena exercises, paramilitaries gave the city's 34,000 residents a realistic preview of martial law, stopping traffic after midnight and staging simulated battles near power facilities at dawn. Lt. Col. Linas Idzelis, the LRU's commander, frames the mission starkly: "Everybody needs to know their role in a formation and have the skills to fight."[2]

Estonia has developed perhaps the most sophisticated total defense model. The Estonian Defense League, with over 25,000 members organized into territorial defense units, operates alongside cyber defense organizations and women's defense leagues. Estonian defense planning assumes that professional forces would conduct mobile defense while territorial units delay enemy advances, protect critical infrastructure, and maintain government continuity.[3]

More importantly, Estonia has integrated cyber defense into national strategy with unprecedented sophistication. Following devastating cyberattacks in 2007 that paralyzed government, banking, and media systems, Estonia established the NATO Cooperative Cyber Defence Centre of Excellence in Tallinn. Estonian doctrine treats cyber as coequal with conventional domains, recognizing that modern conflict begins in digital space long before kinetic operations commence.[4]

Latvia faces unique challenges with a significant Russian-speaking minority comprising roughly one-quarter of its population. Latvian defense strategy emphasizes social cohesion alongside military capability, working to integrate all communities into national defense frameworks. The Latvian National Guard has expanded significantly, while defense education programs target youth across all linguistic communities.[5]

What unites Baltic approaches is rejection of the idea that small nations must remain vulnerable. Baltic forces wouldn't attempt to defeat a Russian invasion independently for that's NATO's collective responsibility. Instead, they focus on making aggression prohibitively costly during the critical 48-72 hours before alliance reinforcements arrive. Every bridge prepared for demolition, every territorial defense unit ready to harass advancing forces, every civilian trained in resistance operations raises the price of aggression.

THE NORDIC EVOLUTION: FROM NEUTRALITY TO NATO INTEGRATION

The Nordic transformation may prove even more consequential than Baltic readiness. Sweden and Finland's NATO accession fundamentally altered European security architecture, adding sophisticated militaries and strategic depth to the alliance's northern flank while abandoning neutrality policies that defined their Cold War identities.

The Finnish Case

Finland's decision to join NATO in 2023, driven directly by Russia's invasion of Ukraine, brought the alliance an 830-mile border with Russia and a military culture built on defensive excellence. Finnish total defense doctrine, developed during decades of non-alignment, offers NATO a proven model for national resilience.

Finland's Defence Forces today plan for a wartime strength of about 280,000 troops built on universal male conscription and a large trained reserve; the overall reserve pool is roughly 870,000, with moves under way to increase this toward one million by 2031. Rapid mobilization of the 280,000 wartime force, but the "870,000" figure is better understood as the total reserve pool, not the fielded wartime strength. Universal conscription, complemented by voluntary service for women, is indeed central to Finland's deterrent posture and supports the "everyone has a role" culture the passage highlights.[6]

Finland maintains one of the world's most extensive civil defense shelter systems, with about 50,500 designated bomb or civil protection shelters nationwide. These facilities have capacity for roughly 4.8 million people in a country of around 5.5–5.6 million, which justifies the claim that they can protect "nearly" the entire population, though not literally every individual. Law and planning assumptions emphasize readiness within hours to days, and a high share of shelters are hardened and equipped for protection against conventional and CBRN threats.[7]

Finland's "security of supply" system, run through the National Emergency Supply Agency, deliberately maintains strategic reserves of critical goods such as fuels, certain foods, and key medical supplies to sustain society under prolonged disruption. Continuity-of-government and continuity-of-operations planning distribute responsibilities and decision-making across ministries, agencies, and regional authorities, precisely to reduce vulnerability to decapitation or centralized paralysis. These arrangements sit within a broader "total defence" tradition that integrates civilian and military planning rather than treating them as separate silos.[8]

Finns receive recurring guidance on household preparedness, including recommendations on supplies and behavior in crises; public awareness campaigns and drills are routine and feed into a "culture of readiness." Education policy and dedicated media-literacy initiatives explicitly address hybrid threats and disinformation, and Finland consistently ranks at or near the top of European indices measuring resilience to information operations. National defense courses and other civic programs reinforce the idea that comprehensive security rests on contributions from all sectors, agriculture, industry, infrastructure, and the digital economy as much as the armed forces.[9]

In a 2022 article, I summed up the Finnish impact as follows:

For years, there has been little room for argument that Finland is laser focused on how to defend its territory. The Finns have a long history of living with the Russians including a century of being part of the Russian Empire itself (1809-1917). Knowing the Russians as well as they do, they are organized to prepare when necessary to defend their nation against their big neighbor when a crisis erupts.

Unlike others in Europe, Finland never bought into the idea of East-West peace lasting forever following the collapse of the Soviet Union. Indeed, just months after the dissolution of the Soviet Union, the Finns inked a $3 billion agreement to purchase 64 F-18 fighters — a major investment at the time, even as the world seemed ready to embrace a post-Cold War era.[10]

The F-18 procurement was a decision which underwrote enhanced Finnish sovereignty, while making a key tie to NATO interoperability even as Helsinki threaded the needle between Russia and NATO for decades more. That legacy was echoed when Finland announced their decision to buy the F-35 on December 10, 2021; picking the F-35 (and not the offering from close neighbor and militarily neutral partner Sweden) clearly factored in the benefits of working more closely with key strategic allies to deflect Russian leader Vladimir Putin's efforts to go back to a world in which Finlandization was a word.

While Finland's official entry into NATO still has a way to go, alliance members have already begun planning how to integrate Finnish airpower into its strategies to counter Russia.[11] There is a base of knowledge to work from: Finland has been working cross-border airpower training for a number of years with Norway and Sweden. Now Finland will be fully integrated with the other F-35 partners in the region — Norway, Denmark, Poland the Netherlands and Belgium — along with other F-35 operators in the US, UK and Germany.[12]

This means that when the Finns fly their aircraft, they will be part of a significant Intelligence, Surveillance and Reconnaissance (ISR) belt looking deep into areas of Russian interest and can provide C2 links to create a more integrated force response, dependent upon national decisions.

As I have written for years, the F-35 is NOT a traditional fighter aircraft; it is a flying combat system whose capabilities become magnified the more F-35s are in the air. (Relatedly, the US and the European F-35 partners need to move more quickly on working the F-35 as an integrated force and its ability to deliver longer range strike against Russian targets in case of conflict.)

This means that Putin now faces a much more integrated and lethal force which can engage across the spectrum of conflict. Ironically, the Russians have skillfully generated Nordic defense collaboration and much closer working relationships with the Balts and Poland as well.

Given its location, Finland is really a key state affecting how the Russians play the geopolitical game along its border with the NATO nations. My travels to Finland and continuing discussions with Finns have taught me much about how they look at the evolving strategic situation.

Their perspective and approach to defense modernization was well articulated by Jukka Juusti, Permanent Secretary in the Ministry of Defence, in an interview conducted during a visit to Helsinki in February 2018. Finland is clearly focused on mobilization and security of supply as key foundations to national defense. And as military transformation unfolds, these core capabilities become increasingly important to deal with the core challenge identified in the Finnish defense policy document published in 2017: "The threshold for the use of military force is lower and the time to respond shorter."

According to Juusti: "If you look at the map of Finland, it's not an island but in practice we are an island. The vast majority of our trade is coming by ships. In that sense we are an island and this means that we have taken the security of supply always very seriously. It is the nature of Finland that we believe that we have to be able to take care of some of the most vital things by ourselves.

"That's the reason, for example, that security of supply is so important for us. For example, with regard to ammunition and those kinds of supplies, we have a lot of stocks here in Finland. Of course, with regard to some of the equipment we never can have enough in our own resources.

"The security of supply has got another respect also, which is the civilian side of the aspect. We have a security of supply agency, which is extremely important for us and it takes care of the civilian part of the security of supply. For example, electricity and telecommunications are vital for the survival of the nation, and one needs have to have the security of supply in those areas.

"Security of supply agency collects the money in such a way that they are financially safeguarded. Whenever we buy some gasoline, they collect some part of that purchase for the security of supply funds. It is organized in that way. We are continuously investing, in effect, in security of supply for the civilian sector.

"And we think broadly about civilian defense as part of our mobilization strategy. That's the reason we were still building shelters for the civilians, both to maintain infrastructure in times of crisis and for civilian protection as well."[13]

Holding such a perspective means that Finns are hardly be shocked by current Russian actions and behavior.

But Finland coming into NATO, given such a realistic Finnish view of the nature of the defense challenge, should push other NATO nations including the United States to get realistic about the depth of the defense challenge posed by Russia and China and how significant the changes need to be to shape a realistic defense approach going forward.[14]

The Swedish Case

Sweden's strategic transformation is striking. A state that practiced armed non-alignment throughout the Cold War has now entered NATO as a frontline contributor, bringing advanced military technology, a sophisticated defense-industrial base, and geography that helps turn the Baltic Sea into a more defensible, NATO-dominated maritime space.

Swedish defense spending has risen sharply in the last decade, reversing post–Cold War cutbacks and moving from roughly 1 percent of GDP toward and beyond the 2 percent NATO benchmark. The 2017 decision to reactivate conscription signaled that Stockholm no longer regarded its neighborhood as benign, while the remilitarization of Gotland has turned the island back into a key position for monitoring and controlling access across the central Baltic.[15]

Sweden's contribution runs well beyond force structure. Its defense industry fields world-class systems—including the JAS 39 Gripen fighter, the Carl Gustaf recoilless rifle, and the Archer artillery system—that are now embedded across allied forces. Decades of operating quiet conventional submarines and surface combatants against Soviet and then Russian activity in the Baltic have produced niche expertise in shallow-water anti-submarine warfare, a skill set now directly integrated into NATO's Baltic ASW exercises and planning.[16]

Sweden's support for Ukraine has been extraordinary in both scale and sophistication. By 2025, Swedish assistance totaled approximately SEK 80 billion (roughly $8 billion) since 2022, with SEK 29.5 billion allocated for 2025 alone. This represents a massive commitment for a country of 10 million people, placing Sweden among the largest per-capita contributors to Ukrainian defense globally. The commitment extends well into the future, with Sweden pledging $3.6 billion annually for 2026-2027, supplemented by $900 million in yearly civilian funding through 2028.[17]

The scope of Swedish assistance extends far beyond financial contributions. Sweden has provided cutting-edge military systems, including the highly effective Archer self-propelled artillery systems. Ukraine now operates 44 Archer systems, with 18 additional units included in Sweden's latest aid package. Swedish Defense Minister Pål Jonson has personally confirmed the effectiveness of these systems against Russian targets during visits to Ukrainian units. Sweden's most recent aid package, announced in September 2025, totaled $836 million and included what Swedish officials described as "Swedish-Ukrainian surprises" for the battlefield, artillery systems, marine defense capabilities, radar systems, and undisclosed advanced technologies.[18]

The Swedish approach emphasizes quality and sustainability rather than mere quantity. Aid packages include not only weapons systems but comprehensive support packages with maintenance solutions, spare parts, ammunition, and technical support. This holistic approach ensures that donated equipment remains operational long-term, maximizing its impact on Ukrainian defense capabilities.

As NATO Deputy Secretary General Radmila Shekerinska observed, Sweden has become "an outstanding supporter of Ukraine, providing billions of euros in military assistance," including "tanks, armoured vehicles, missiles and artillery" that are "helping Ukraine fight for its freedom."[19]

The Norwegian Case

Norway, a founding NATO member, has amplified its strategic importance. Norwegian forces provide critical Arctic expertise as climate change and Russian ambitions make the High North increasingly contested. The Norwegian Joint Headquarters at Bodø coordinates allied operations in the region, while Norwegian intelligence monitors Russian Northern Fleet activities from installations along their shared border.

Norway's reaction to the Trump administration's shifting Ukraine policy in its second term illustrates how a European middle power can move quickly to offset perceived gaps in great-power support. In early March 2025, amid a suspension of U.S. military aid and intelligence sharing to Kyiv, Norwegian parliamentary leaders agreed to raise their planned 2025 support for Ukraine by 50 billion Norwegian kroner (about 4.6 billion dollars), bringing the total to roughly 85 billion kroner (about 7.8 billion dollars) and more than doubling the level previously agreed in late 2024.[20]

Prime Minister Jonas Gahr Støre framed the expanded package as both a response to Ukraine's urgent battlefield needs and a signal of European commitment at a time when "changes in the United States have weakened support for Ukraine," a situation he described as "very sad." Støre argued that the additional funds were intended to strengthen Ukraine's position, support a viable peace, and enhance European security, explicitly linking Norway's move to broader concerns about the direction of U.S. policy under President Trump.[21]

The increase was channelled through the Nansen Support Programme for Ukraine, first launched in early 2023 with broad cross-party backing and later expanded so that its total framework reached about 205 billion kroner for 2023–2030. Analyses of Ukraine aid relative to economic size indicate that this long-term framework corresponds to roughly 0.5–0.6 percent of Norway's GDP and places Norway among the world's leading contributors in GDP-adjusted terms, underpinned by an unusually strong consensus in the Storting for sustained support.[22]

As part of this broader effort, Norway worked closely with other European allies on specific air-defence capabilities, including a joint initiative with Germany to provide Patriot systems and associated equipment for Ukraine. Oslo committed about 7 billion kroner (around 695–700 million dollars) for air-defence systems sourced from German and Norwegian industry, an arrangement intended both to accelerate deliveries and to demonstrate European solidarity in meeting Ukraine's most acute defence requirements.[23]

Despite criticism in Europe of Trump's decisions on Ukraine aid, Norway maintained visible high-level engagement with Washington. In late April 2025, Støre met President Trump at the White House and publicly stressed that Norway had "tripled" its military support to

Ukraine and would allocate close to 8 billion dollars that year, presenting this as a complement to diplomatic efforts to secure a just peace. Støre cast Norway's expanded contribution as a way to strengthen Ukraine's position "on the battlefield and at the negotiating table," while signalling that robust European action could help stabilize transatlantic strategy even amid U.S. policy volatility.[24]

The Danish Case

Denmark is pursuing an ambitious defense modernization program that spans from F-35 fighter jet acquisitions to coordinating European support for Ukraine's air defenses, while simultaneously bolstering its Arctic presence in response to growing geopolitical pressures.

Danish Defense Minister Troels Lund Poulsen has confirmed plans to significantly expand the country's F-35 Lightning II fleet beyond the original 27-aircraft order. Speaking at a Center for Strategic and International Studies event in July 2025, Poulsen indicated that purchasing additional F-35 combat aircraft "will be the way forward," with at least 10 additional jets planned for acquisition.[25]

This expansion would bring Denmark's total F-35 fleet to approximately 40 aircraft, representing what Poulsen described as "a very large amount of billions" in investment. The Royal Danish Air Force currently operates 15 delivered F-35As at Skrydstrup Air Base, with six aircraft remaining in the United States for training purposes.[26]

A significant milestone for 2025 will see F-35s begin to occasionally replace aging F-16 fighters in air policing duties, marking Denmark's transition to fifth-generation air power. The country plans to completely phase out its F-16 fleet by 2027, as these aircraft have been in service since the 1980s.[27]

Denmark is also exploring strategic deployments of its F-35 fleet, with plans to upgrade Kangerlussuaq airport in western Greenland to support F-35 operations. This development would represent one of the most significant improvements to Greenland's security infrastructure in decades.[28]

Denmark has emerged as a key coordinator in European efforts to provide advanced air defense systems to Ukraine. Poulsen secured an agreement on behalf of European nations to purchase and donate 10 Patriot air defense systems to Ukraine, marking a breakthrough in addressing critical capability gaps.

"The most important part right now is that Europe would be able to buy military equipment here in the U.S., so we can donate these military systems directly to Ukraine," Poulsen explained at the CSIS event. This arrangement addresses previous constraints where no additional Patriot systems were available for immediate purchase from U.S. manufacturers.

Denmark has dramatically increased its Arctic defense spending in response to growing regional tensions and renewed U.S. interest in Greenland. In January 2025, Copenhagen announced a $2.05 billion investment package to strengthen security in the Arctic and North Atlantic regions.[29]

The comprehensive package includes funding for three new Arctic naval vessels capable of carrying helicopters and drones, long-range surveillance drones with advanced imaging capabilities, and enhanced satellite coverage. This investment follows a separate $1.5 billion Greenland defense package announced in December 2024, covering inspection ships, surveillance drones, and additional dog sled patrol teams.

"We must face the fact that there are serious challenges regarding security and defense in the Arctic and North Atlantic," Poulsen stated. "For this reason, we must strengthen our presence in the region".[30]

Currently, Denmark's Arctic capabilities consist of four aging inspection vessels, one surveillance aircraft, and 12 dog sled patrols monitoring an area four times the size of France. The new investments aim to address what has been characterized as a "security black hole" in the strategically vital region.[31]

These defense enhancements come amid President Donald Trump's repeated assertions that U.S. control of Greenland is "an absolute necessity" for American national security. Danish and Greenlandic officials have firmly rejected any notion of transferring sovereignty, with Greenland Prime Minister Múte Egede stating that "Greenland belongs to the people of Greenland".[32]

Rather than acquiring expensive maritime patrol aircraft, Denmark has opted for a cost-effective approach by leasing flight hours from Norway's P-8 Poseidon fleet. This arrangement provides surveillance capacity over the Arctic without the substantial investment required for independent aircraft procurement.

"Right now, we are looking into buying [flight] hours from the P-8 system... for capacity," Poulsen explained. "They have [the aircraft] in Norway and Germany and [we do] not [have interest in] buying it ourselves, but buying hours" alone.

This decision reflects Denmark's pragmatic approach to defense procurement, recognizing the benefits of multinational cooperation in addressing capability requirements while managing costs effectively.

Denmark's defense initiatives align with broader NATO requirements and changing security dynamics. The alliance has set new targets of 5% of GDP for defense spending, creating pressure for increased military investment across member states. Denmark committed over 3% of its GDP, approximately $7 billion, to defense spending over two years in early 2025.

The country's defense strategy emphasizes interoperability with NATO allies and standardization around proven platforms like the F-35. As Poulsen noted, Denmark cannot afford to operate multiple types of fighter aircraft due to its size and resource constraints, justifying the focus on a standardized F-35 fleet.

These developments demonstrate Denmark's commitment to addressing multiple security challenges simultaneously: supporting Ukraine's defense against Russian aggression, strengthening Arctic sovereignty, and modernizing its own military capabilities.

The coordinated approach reflects both the interconnected nature of contemporary security challenges and Denmark's role as a reliable ally in addressing them.

Danish control of Greenland and the Faroe Islands gives NATO strategic positions in the North Atlantic, while Danish forces contribute disproportionately to allied operations. The Danish commitment to defense spending increases reflects awareness that small nations must demonstrate serious commitment to collective defense.

THE NORDIC SUPPORT EFFORTS FOR UKRAINE

Nordic countries have pioneered innovative multilateral approaches to supporting Ukraine. In August 2025, Sweden, Denmark, and Norway jointly funded a $500 million package of U.S. military equipment for Ukraine under NATO's new Prioritised Ukraine Requirements List (PURL) initiative. Sweden contributed $275 million to this coordinated effort, demonstrating commitment to multilateral approaches. NATO Secretary-General Mark Rutte commended the Nordic countries for their "fast action and steadfast support for Ukraine," noting that "NATO members provide 99% of military support to Ukraine."[33]

The PURL initiative was explicitly designed to "put Ukraine in the strongest possible position as peace efforts, led by President Trump and his administration, continue." This framing illustrates how Nordic countries have sought to position their support as complementary to rather than contradictory to American diplomatic efforts, maintaining alliance unity even amid policy disagreements.[34]

The Nordic approach emphasizes not merely providing equipment but building sustainable capabilities. Aid packages include comprehensive maintenance solutions, training programs, spare parts, and technical support. This focus on sustainability reflects lessons learned from earlier assistance efforts where equipment became inoperable due to lack of maintenance capacity or spare parts availability.[35]

POLAND: THE EASTERN ANCHOR

If the Baltics provide forward defense and the Nordics add strategic depth, Poland serves as NATO's eastern anchor—the alliance's largest and most powerful eastern member, positioned between German industrial heartland and Ukrainian battlespace.

Poland's defense transformation has been nothing short of revolutionary. The nation has committed to fielding the largest land force in Europe outside Russia, with plans to expand from roughly 150,000 to 300,000 active personnel. Defense spending has surged past 4% of GDP, double NATO's 2% target and higher than any other alliance member. This isn't rhetorical commitment but actual procurement: Poland has ordered over 1,000 K2 Black Panther tanks from South Korea, 500 HIMARS rocket systems from the United States, 48 FA-50 fighter aircraft, and dozens of Apache helicopters.[36]

This military expansion reflects strategic calculation. Polish defense planners recognize their nation would serve as the primary logistics hub for any NATO operation defending the Baltics or supporting Ukraine. Poland has also become a leading voice for robust NATO posture toward Russia. Polish diplomacy consistently advocates for enhanced forward presence, arguing that deterrence requires visible capability positioned where it matters. The nation hosts NATO's Enhanced Forward Presence battlegroup and advocates for permanent basing of substantial allied forces on Polish territory.

Poland is building a highly layered, missile-centric defense posture that combines Patriot for air and missile defense with HIMARS for long-range precision fires, driven by expectations that any future conflict will feature intensive missile and drone use.

Poland's Wisła program centers on the U.S. Patriot system, with an initial two Patriot Configuration-3+ batteries already contracted and a second, much larger phase cleared by the U.S. for up to a roughly $15 billion integrated air and missile defense package, including additional launchers, radars and the IBCS battle management network. Recent support deals worth about $2 billion and local launcher production contracts indicate that Warsaw is not only buying Patriot batteries but also investing in training, logistics and domestic industrial participation to keep the system ready and scalable over time.[37]

On the land fires side, Poland has moved to acquire HIMARS on an unprecedented scale, with a framework agreement to buy 486 launcher-loader module kits for M142 HIMARS and associated rockets and ATACMS-class missiles, far beyond its initial 20-system order. Follow-on negotiations under the Homar-A program are focused on integrating HIMARS with Polish chassis, fire-control systems and local munitions production, turning the capability into both a deterrent and a domestically supported long-range strike complex.[38]

These investments are framed explicitly against a perceived Russian missile, rocket and drone threat from Kaliningrad, Belarus and over the Ukrainian battlespace, where Polish analysts and officials see massed salvos and swarming unmanned systems as central features of any future conflict. Broader European debates on a "drone crisis" and a regional "drone wall" underscore that Warsaw expects attacks not only on frontline forces but also on critical infrastructure, logistics hubs like Rzeszów, and urban centers, driving the push for dense, layered defenses rather than isolated point systems.[39]

Perhaps most significantly, Poland has embraced its role as Ukraine's most steadfast supporter. Polish territory serves as the primary logistics corridor for Western military aid. Polish trainers work with Ukrainian forces. Polish diplomacy consistently pushes for stronger Western support.

This isn't altruism but strategic awareness: Polish security depends fundamentally on Ukrainian success. A Russian victory in Ukraine would place hostile forces along extended sections of Polish border.

THE HYBRID WARFARE FRONTIER

One area where these nations truly lead NATO is confronting hybrid threats, the gray zone operations Russia employs to destabilize, confuse, and weaken adversaries without triggering Article 5 collective defense obligations.

Lithuania faces nearly 600 balloons carrying contraband from Belarus this year alone, triple 2024 levels. These forced multiple Vilnius airport closures, disrupting thousands of passengers. Lithuanian officials characterize this as deliberate hybrid attack aimed at destabilization. The banality of the method, balloons carrying cigarettes, shouldn't obscure its strategic purpose: demonstrating vulnerability, imposing costs, testing response capabilities, and normalizing boundary violations.[40]

Finland confronts a spectrum of hybrid pressures from Russia. In late 2023, Russian authorities were widely accused of facilitating the movement of Middle Eastern and African migrants toward the Finnish border, turning humanitarian flows into a tool of coercion. Finland responded by progressively closing its eastern land border crossings and advancing emergency legislation to handle what it defined as 'instrumentalized' or weaponized migration, steps that provoked domestic legal and human rights debate but also signaled political resolve and alliance cohesion.[41]

Baltic states face persistent cyberattacks, disinformation campaigns, and GPS interference. Russian forces regularly jam GPS signals near Baltic borders, affecting civilian aviation and maritime navigation. Disinformation campaigns amplify social divisions, promote conspiracy theories, and undermine confidence in democratic institutions.

Estonia is widely regarded as a pioneer of e-governance, with nearly all public services digitized and underpinned by strong cybersecurity, redundancy of critical data (including "data embassy" backups abroad), and continuity planning to maintain operations under attack.[42]

Its model emphasizes distributed architecture, secure digital ID, and regular cyber exercises, which together enhance resilience against both state and criminal cyber threats.

Finland integrates media literacy, critical thinking, and digital skills throughout its national curriculum from early grades onward and supports this with teacher training and nationwide programs involving schools, libraries, media, and civil society.[43]

This whole-of-society approach is frequently cited as a key factor in Finland's resilience to disinformation compared with most European states.

Lithuanian intelligence and security services publicly highlight and document Russian espionage, sabotage, and information operations, including expulsions of Russian "diplomats" identified as intelligence officers and court cases against recruited agents.[44]

Regular public reports and media collaborations aim to deter recruitment, raise societal awareness, and signal that Russian activities are being monitored and exposed.

Poland has created Cyberspace Defence Forces tasked with securing military networks and conducting the full spectrum of cyber operations, including defensive, reconnaissance, and offensive actions.

Polish strategy and official statements stress both protection of national and allied systems and the development of indigenous offensive tools and training ranges to operate effectively in wartime cyberspace.[45]

The experience these nations gain confronting hybrid threats provides NATO invaluable lessons. Western European allies face similar challenges, Russian disinformation campaigns throughout Europe, cyberattacks on critical infrastructure, energy coercion, but often lack the urgency and experience that proximity to Russia provides. Baltic, Nordic, and Polish expertise increasingly informs alliance-wide approaches to hybrid defense.

POLAND'S OPERATION HORIZON

In late November 2025, Poland launched one of its largest peacetime security operations, deploying up to 10,000 troops across the country to protect vital infrastructure from an escalating campaign of sabotage. Operation Horizon, announced jointly by Defense Minister Władysław Kosiniak-Kamysz and Interior Minister Marcin Kierwiński on November 19, represents a significant escalation in Poland's response to hybrid warfare tactics attributed to Russian intelligence operations.[46]

The catalyst for this massive deployment was a railway explosion near Warsaw on November 16, which damaged a critical line used for transporting aid to Ukraine. Prime Minister Donald Tusk characterized the incident as an "unprecedented act of sabotage," marking a dangerous new phase in attacks targeting Poland's strategic infrastructure. Polish authorities identified two Ukrainian citizens, allegedly recruited by Russian intelligence services, as the primary suspects. The pair fled through the Terespol crossing into Belarus immediately after the attack, highlighting the complex transnational nature of modern hybrid threats.[47]

Operation Horizon mobilizes forces from all branches of the Polish Armed Forces, including Territorial Defense Forces, special operations units, and cybersecurity teams. These troops work in coordination with police, border guards, the Internal Security Agency, and counterintelligence services under a unified command structure.

The operation focuses on securing railways, airports, bridges, viaducts, energy facilities, and telecommunications networks, the essential arteries of Poland's economy and its role as a critical logistics hub for supporting Ukraine.

The military deployment employs advanced surveillance capabilities, including reconnaissance drones and helicopters, to monitor vulnerable sections of infrastructure. Within days of the operation's launch, patrols had been established at 80 critical rail sections across Poland, with forces concentrated in areas deemed at highest risk. This visible military presence serves a dual purpose: providing actual security while also creating a deterrent effect by introducing "additional risks" for potential saboteurs, as Defense Minister Kosiniak-Kamysz explained.[48]

Beyond immediate security measures, Operation Horizon reflects Poland's broader strategic adaptation to hybrid warfare. The operation aims not only to protect physical assets but also to counter disinformation campaigns and bolster public confidence in the face of destabilization efforts. To enhance civilian participation in national security, Poland plans to introduce a mobile application that will enable citizens to report suspicious activities, particularly during vulnerable periods such as the holiday season.

The military deployment employs advanced surveillance capabilities, including reconnaissance drones and helicopters, to monitor vulnerable sections of infrastructure. Within days of the operation's launch, patrols had been established at 80 critical rail sections across Poland, with forces concentrated in areas deemed at highest risk. This visible military presence serves a dual purpose: providing actual security while also creating a deterrent effect by introducing "additional risks" for potential saboteurs, as Defense Minister Kosiniak-Kamysz explained.

Beyond immediate security measures, Operation Horizon reflects Poland's broader strategic adaptation to hybrid warfare. The operation aims not only to protect physical assets but also to counter disinformation campaigns and bolster public confidence in the face of destabilization efforts. To enhance civilian participation in national security, Poland is introducing a mobile application that will enable citizens to report suspicious activities, particularly during vulnerable periods such as the holiday season.

THE INFRASTRUCTURE OF DETERRENCE

These nations understand that deterrence requires visible investment in capability to receive, sustain, and employ allied reinforcements. They've invested accordingly.

Poland, the Baltic states, Norway, and Finland have all invested heavily in rail and port infrastructure with an explicit eye to enabling rapid NATO reinforcement, not just commercial traffic. These upgrades increase throughput, redundancy, and options for moving heavy forces into the Baltic–Nordic theater, making the adage "reinforcements delayed are reinforcements denied" a practical planning principle rather than a rhetorical flourish.[49]

Poland has positioned itself as the main continental land bridge for NATO into the eastern flank, expanding and modernizing rail and road links from Germany toward the Baltic states and Ukraine, including dual-use corridors designed to carry heavy armor and large troop movements. In parallel, Lithuania, Latvia, and Estonia are pursuing Rail Baltica to create a standard-gauge, high-capacity link from Poland to Tallinn, which is routinely framed in NATO and EU debates as a military mobility project as much as a commercial one.

Poland has expanded port capacity at facilities such as Gdynia and Gdańsk to handle more roll-on/roll-off traffic and project cargo, which directly supports reception of allied heavy equipment and logistics flows into the eastern flank. Baltic ports in Lithuania, Latvia, and Estonia are being upgraded not only for energy and container traffic but also for hosting allied naval task groups and supporting maritime domain awareness, reflecting the shift of the Baltic Sea into what analysts increasingly describe as a "NATO lake."[50]

Norway continues to treat the High North as a priority, refurbishing and expanding bases in northern regions to host larger and more frequent allied contingents, with recent agreements giving the United States and other allies broader access to airfields and pre-positioning sites. New initiatives such as the amphibious warfare center at Sørreisa illustrate a deliberate effort to create infrastructure that can receive, house, and sustain substantial NATO reinforcement packages for Arctic and North Atlantic operations.[51]

Finland's long coastline on the Gulf of Finland and Gulf of Bothnia, along with its existing commercial port network, offers NATO additional entry points and staging areas for maritime logistics and coastal defense systems. As Finland integrates into NATO planning, analysts highlight the potential to link Finnish ports and coastal defenses with Swedish and Baltic capabilities to form a layered structure that both supports reinforcements and constrains Russian movement in the Baltic approaches.

Across these investments, the strategic logic is consistent: infrastructure that cannot move heavy forces quickly and at scale under crisis conditions effectively nullifies paper commitments to reinforce the region. Enhancing rail lines, ports, and bases reduces deployment times, complicates Russian planning in the Baltic–Nordic theater, and turns military mobility into a tangible deterrent rather than a vulnerable assumption.

These investments reflect lessons from Ukraine. The war has demonstrated that modern conflict consumes ammunition and materiel at rates Western militaries haven't anticipated. Logistics matters as much as combat capability. Supply lines determine operational possibility. Nations investing in infrastructure that allows allies to move, supply, and sustain forces contribute as much to deterrence as those fielding additional brigades.

SEVERING THE SOVIET RAIL LINK

Latvia's emerging plan to dismantle rail links to Russia and the parallel build-out of Rail Baltica together mark a strategic shift that severs Soviet-era dependencies and hard-wires the Baltic states into NATO's logistics network. This dual process blends economic sacrifice, military prudence, and geopolitical signaling into a long-term reorientation of regional infrastructure from Moscow to Western Europe.[52]

The railways of Estonia, Latvia, and Lithuania still use the 1,520 mm "Russian" broad gauge inherited from the Russian Empire and Soviet Union, whereas most of Europe uses the 1,435 mm standard gauge. This incompatibility means that key Baltic transport arteries physically align with Russia and Belarus rather than with core EU and NATO partners, raising concerns that infrastructure could facilitate hostile military movements as much as Allied reinforcement.[53]

Latvia's leadership has confirmed that the government will evaluate dismantling railway sections leading to Russia, with an assessment to be prepared by the end of the year in coordination with the National Armed Forces and Baltic partners. President Edgars Rinkēvičs and Prime Minister Evika Siliņa have framed the measure as one of several options to strengthen national defense along a tense eastern border, with final decisions contingent on security and socio-economic analysis.[54]

Rail lines remain central to modern military logistics, as illustrated by Russia's extensive reliance on rail to sustain operations in Ukraine. By breaking direct rail connectivity, Latvia would force any aggressor to depend on slower and more exposed road convoys or invest time and resources in restoring track, thereby improving NATO's prospects for delaying and disrupting a cross-border attack.

Baltic transit volumes from Russia have already declined sharply in recent years as Moscow has redirected cargo away from Baltic ports, but remaining flows, especially bulk commodities such as grain, still account for a significant share of Latvia's rail business. Fully dismantling cross-border lines would likely eliminate what is left of this trade, embedding the security choice in long-term economic losses and making reversal politically and commercially difficult.

Rail Baltica, a new 870-kilometer standard-gauge corridor linking Estonia, Latvia, and Lithuania with Poland, is designed to connect the Baltic states directly to the European rail network. The project, now slated for completion around 2030 after delays and cost growth, has evolved from a market-integration initiative into a flagship of NATO-compatible military mobility in the region.[55]

Using standard gauge, Rail Baltica would allow NATO trains to move troops and heavy equipment from Western Europe to the Baltics without time-consuming gauge-change procedures or large-scale road movements. Project documents highlight the potential to replace very long military convoys with single trains and to develop dual-use terminals such as expanded facilities near Kaunas that can handle both commercial and military cargo flows.[56]

Taken together, dismantling Soviet-gauge links to Russia and building a NATO-standard north-south axis amount to a gray-zone operation conducted through infrastructure rather than armed force. The permanence of track gauge and route alignment effectively locks in the Baltic states' westward orientation for decades, signaling that their rail systems will serve Alliance reinforcement and EU integration, not renewed dependence on Moscow.

If current plans hold, by around 2030 the Baltic states will operate a core rail backbone tied into Poland and Western Europe while progressively reducing or re-purposing Soviet-gauge connections to Russia and Belarus. This transformation, implemented with excavators, bridges, and new track rather than combat formations, could prove as strategically consequential for deterrence and defense as more traditional military posture decisions.

CULTURAL FACTORS: MEMORY AND COMMITMENT

Understanding why these nations lead NATO requires acknowledging cultural and historical factors that Western European allies don't share. Every Baltic state experienced Soviet occupation. Every Finn remembers the Winter War. Every Pole knows the history of partition. These aren't distant memories but lived experience, grandparents' stories, family trauma, national narratives that shape identity.

This produces fundamentally different risk calculus. When Western European politicians debate defense spending increases, they weigh opportunity costs, education, healthcare, infrastructure competing for resources. When Baltic, Nordic, and Polish leaders make similar decisions, they assess existential questions: would we survive Russian aggression, could we maintain sovereignty, what price independence?

Public opinion reflects these differences. Polling consistently shows Baltic, Nordic, and Polish populations supporting defense spending increases and NATO membership at much higher rates than Western European counterparts. These aren't governments imposing unwanted policies but responding to genuine public concern.

CHALLENGES AND LIMITATIONS

Small populations limit mobilization potential: the entire Baltic population combined barely exceeds that of metropolitan Paris. Economic constraints affect sustainability of current spending levels, particularly for Poland where defense expenditures compete with pressing domestic needs. Geographic vulnerability remains despite all preparations, the Suwalki Gap represents genuine strategic weakness, Finnish border length with Russia creates extended defense requirements, Baltic states remain potentially isolated if Russia acts decisively.

Dependence on allied reinforcement creates political vulnerabilities. These nations' strategies assume NATO Article 5 solidarity that American, British, French, German, and other forces would actually deploy to defend them. Recent political developments in the United States and some European nations raise questions about reliability of these commitments.

Hybrid threats exploit democratic vulnerabilities that even sophisticated countermeasures struggle to address. Disinformation campaigns succeed not through crude propaganda but by amplifying genuine social divisions. Migration weaponization forces democracies to choose between humanitarian values and border security. Cyber operations target critical infrastructure that interconnected societies depend upon. Defending against these threats without undermining democratic norms and civil liberties presents genuine dilemmas.

IMPLICATIONS FOR NATO

The innovations these nations pioneer increasingly influence NATO strategy. The alliance's enhanced forward presence reflects Baltic advocacy for visible deterrence. NATO's attention to military mobility follows from Eastern European emphasis on rapid reinforcement. Growing focus on hybrid threats acknowledges challenges these nations confront daily. Total defense concepts enter alliance doctrine after decades of Nordic development.

Most fundamentally, these nations remind NATO that deterrence requires credibility, visible capability, demonstrated commitment, political will to actually employ force if necessary. Western European allies may debate force structure and procurement timelines, but Baltic, Nordic, and Polish actions demonstrate what serious defense commitment entails.

This leadership carries risks. If these nations press for NATO postures that other allies consider provocative, internal alliance tensions could emerge. If they advocate for policies toward Russia that Western European nations consider too confrontational, diplomatic complications arise. Balancing appropriate deterrence with avoiding unnecessary escalation requires careful calibration.

Yet the fundamental contribution these nations provide, taking their own defense seriously, investing substantially in capability, developing innovative approaches to emerging threats, demonstrating political will to defend sovereignty, strengthens NATO immeasurably. Alliances work when members contribute according to capability and when those most exposed to threat lead in confronting it.

CONCLUSION

The Baltic states, Nordic nations, and Poland haven't simply increased defense spending or expanded force structure. They've fundamentally rethought what national defense requires in an era of hybrid warfare, renewed major power competition, and threats to the rules-based international order. Their innovations in total defense, civil resilience, cyber security, and conventional capability provide NATO models for how democracies can defend themselves without abandoning democratic values.

When Lithuanian paramilitaries conduct urban warfare drills, when Finnish citizens check their household crisis preparedness, when Polish forces field modern equipment at scales not seen in

Europe for decades, when Nordic nations abandon neutrality to join collective defense, they demonstrate that free societies can match authoritarian threats when citizens understand what's at stake.

These nations lead not from power but from necessity, not from aggressive intent but from defensive imperative. Their message to the alliance is clear: deterrence works when potential aggressors believe defense will succeed, when societies demonstrate cohesion, when military capability backs diplomatic resolve.

In an era of renewed Russian aggression, this leadership may prove essential to European security and NATO's continued relevance. The question facing Western European allies isn't whether to follow this lead but whether they'll recognize its necessity before events force the choice upon them.

CHAPTER 4
DIVIDED ALLEGIANCES

Three years into Russia's full-scale invasion of Ukraine, the former Warsaw Pact members of the European Union and NATO find themselves profoundly divided over how to respond to the greatest security crisis in Europe since World War II.

These eight nations, Poland, the Czech Republic, Slovakia, Hungary, Romania, Bulgaria, Estonia, Latvia, and Lithuania, share a common history of Soviet domination and a collective memory of Russian imperialism.

Yet their responses to the Ukraine war have varied dramatically, ranging from unwavering support to active obstruction of Western aid efforts. This divergence reflects not merely different threat perceptions, but fundamental disagreements about sovereignty, European integration, and the nature of the post-Cold War order.

As of October 2025, these divisions have crystallized into distinct groups that undercut European cohesion in shaping a common response to the ongoing war and Europe's dealings with Russia.

I would argue that perhaps former Warsaw Pact states will have a significant impact on Russia's future after Putin, as Putin is in the process of destroying the Russian empire, and leaving behind a very uncertain identity for Russia in what will be a post-Putin world.

THE VANGUARD: POLAND
AND THE BALTIC STATES

The most steadfast supporters of Ukraine among the former Warsaw Pact states have been Poland and the three Baltic republics, Estonia, Latvia, and Lithuania. These countries view Russia's aggression against Ukraine not as a distant conflict, but as an existential threat to their own security. Their geographic proximity to Russia, combined with living memory of Soviet occupation, has shaped an uncompromising stance toward Moscow's imperial ambitions.

Poland has emerged as Ukraine's single most important supporter among European nations when measured by the percentage of GDP committed to aid. According to Polish government sources, Poland's total assistance of PLN 15 billion (approximately €3.3 billion) was among the first significant aid to reach the frontline. This commitment extends beyond financial metrics to encompass military hardware, humanitarian assistance, and refugee support on an unprecedented scale.[1]

The military dimension of Polish support has been particularly significant. Poland has provided Ukraine with substantial quantities of weapons, including MiG-29 fighter jets, T-72 tanks, and over 100 million rounds of ammunition of various types and calibers.

Polish President Andrzej Duda has been vocal about the strategic importance of supporting Ukraine, emphasizing that such support is essential for regional security. He stated: "I care about the security of Ukraine because it is our neighbour and today it stops potential Russian aggression against other countries. By fighting Russia, by blocking Russian imperialism, Ukraine today defends Europe, including us."

Furthermore, Duda has highlighted Poland's proactive role in the collective defense and security cooperation within Europe, particularly within the Bucharest Nine format, and underlined the importance of unity in countering Russian hybrid attacks and full-scale aggression against Ukraine. At the United Nations in April 2025, he reaffirmed Poland's support for Ukraine, condemning the blatant violation of international law by Russia and stressing the importance of collective action to uphold sovereignty and security in the region.[2]

Later, in July 2025, Duda criticized both Ukraine and some of its Western allies for underestimating Poland's contributions to the war effort, asserting that key logistics and infrastructure are treated as if they belong to others, not Poland. In an address marking Poland's Independence Day, he stressed that Poland is at the forefront of aid to Ukraine, both military and humanitarian, because of their shared love of freedom and commitment to oppose tyranny.[3]

These statements reflect Duda's consistent view that supporting Ukraine is vital not only for Ukraine's sovereignty but also for regional and European security, and that collective, international action backed by strong alliances like NATO and the United States is indispensable in countering Russian aggression.[4]

Beyond military support, Poland has assumed a remarkable humanitarian role. According to data from the UN High Commissioner for Refugees and Eurostat, as of mid-2025 Poland hosted approximately 994,000 to 995,000 Ukrainian refugees under temporary protection, representing the largest number in any single European country apart from Germany and the highest absolute total among Ukraine's immediate neighbors.[5]

The three Baltic states, Estonia, Latvia, and Lithuania, have demonstrated perhaps the most remarkable commitment to Ukraine relative to their size and economic capacity. These nations, each with populations under three million, have contributed between 1.5% and 2.2% of their GDP to Ukrainian aid, placing them among the top five contributors globally when measured by this metric.[6]

Estonia has been particularly prominent, providing aid worth 2.2% of its GDP, over four times the proportion of U.S. aid (0.5%). This commitment includes not only financial transfers but also military equipment, particularly anti-tank Javelin missiles and various defensive systems provided in the earliest days of the invasion. Estonian officials have been forthright about the strategic logic behind this support.[7]

As the Estonian Defense Ministry spokesperson explained, the military assistance "has served as a very important political signal of U.S. support" and has "enabled us, among other things, to accelerate critical capability development and to choose U.S. equipment."[8]

Lithuania has institutionalized its support for Ukraine by committing to spend 0.4% of its GDP annually on military support, pledging a total of €200 million through 2026.[9] The Lithuanian Defense Ministry has explicitly stated that "keeping Ukraine safe means taking care of our security," viewing Ukraine's resistance against Russian imperialism as vital for Lithuanian national security.[10]

This commitment extends to technological assistance, with Lithuania becoming one of the first countries to work on Ukraine's reconstruction and investing €21 million directly in Ukraine's arms industry to facilitate joint production initiatives.[11]

Latvia is sustaining one of the highest relative levels of military support for Ukraine among European states, having pledged 0.25% of its GDP for Ukraine aid in 2024, about €112 million, with this commitment already factored into the country's 2025–2026 defense budgets.

Since Russia's invasion, Latvia's total military aid to Ukraine has now reached close to 1% of its GDP, reflecting both continued urgency and public backing for Kyiv. A flagship element of the 2025 package will be the delivery of 42 new Patria 6×6 armored vehicles from Latvian production lines, expected to enhance Ukrainian operational mobility.

In the sphere of drone warfare, Latvia has earmarked €20 million for the international Drone Coalition in 2024, intends to repeat this in 2025, and has separately committed €10 million to direct investment in Ukraine's drone industry, moves repeatedly highlighted in official defense statements and European strategic communications.[12]

The Baltic states' refugee acceptance has been equally impressive relative to their populations. Estonia has hosted approximately 40,000 Ukrainian refugees (about 3% of its population), Latvia between 50,000-53,000 (2.7% of the population), and Lithuania has issued more than 50,000 visas for temporary protection with over 80,000 arrivals claimed (2.8% of the population).[13]

The unwavering support from Poland and the Baltic states reflects a clear-eyed assessment of their geopolitical position. These nations understand that a Russian victory in Ukraine would embolden Moscow to test NATO's resolve elsewhere which they believe would most likely in the Baltic region or Poland.

The determination of these frontline states was underscored by their joint decision to withdraw from the Ottawa Treaty banning landmines which was a step aimed at strengthening their national defenses. Baltic leaders have adopted a proactive stance, pledging to strike military targets inside Russia if attacked, having witnessed Russian atrocities against Ukrainian civilians and determined not to wait for liberation by NATO reinforcements.[14]

THE CZECH REPUBLIC

I did an interview in 2022 with Petr Pavel before he became President of the Czech Republic. In that interview he argued the following:[15] Petr Pavel, a retired four-star general has served as the president of the Czech Republic since March 2023. He is a former Chairman of the NATO Military Committee and previously served as Chief of the General Staff of the Czech Armed Forces. Pavel ran as an independent in the 2023 Czech presidential election on a strongly pro-Western, pro-NATO, and pro-EU platform, emphasizing support for Ukraine and a firm stance toward Russia and China.

> *Pavel underscored that what is unfolding in Ukraine today is a product of the long-range perspectives and policies of President Putin. He noted: "Russia still believes that the very possession of nuclear arsenal makes it a superpower, and enables it to dictate to other countries how to live and how to arrange their international relations."*
>
> *The Russians take no blame for their actions, but argue that the West is the trigger of their actions. He underscored that "They believe that they're perfectly right. All the moves they do are well justified and the vast majority of Russian population stands behind President Putin in this belief.*

For all of us, it means that the situation international and security situation has fundamentally changed.

"Russia is no longer just a strategic competitor. It's a direct threat to our security. President Putin, went so far by declaring a possible use of nuclear weapons against United Kingdom, France and other countries, threatening sovereign countries, including nuclear powers. It's not an issue for us to think if Russia is bluffing or if they are just posing threats, they are determined to use these weapons and they expressed their determination several times pointing to existential threats, but without being very specific."

"When we look at the breadth of Russian strategic thinking, we have to understand that anything that would harm Russian perception of being a global power could be taken as existential threat to Russia and as a justification for using nuclear weapons, tactical or any other level.

"But I believe that any use of nuclear weapons, even smallest tactical nuclear weapon would be breaching the threshold and would fundamentally change the whole paradigm of our security thinking.

"We don't have to necessarily defeat Russia but we face the challenge of dealing with Russian thinking that whatever we do in terms of coming closer to them would be understood as our weakness and sooner or later, we would see them attack another country."

He argued that the war in Ukraine was having a significant impact on a "reshuffling in Europe" whereby the states most directly familiar with Russia and closest to Russia have views different from key states like France and Germany.

He argued that "we see a new, I wouldn't call it fraction line, but potential division within Europe where countries bordering Russia such as Bulgaria, Romania, Baltic countries, Slovakia, Poland, Czech Republic are almost of the same opinion on how to handle the crisis. That means being very tough within the framework of sanctions against Russia, being extremely helpful to Ukraine in terms of humanitarian assistance including military assistance."

The exceptions to this are the states of Hungary and Turkey, as he noted, which form "special cases" as he noted.

He added that one outcome of the war in Ukraine could well be Ukraine becoming part of the European Union, which if this was to happen would enhance the weight of the Russian threat focused countries within the European Union.

He very clearly underscored the importance of working relationships in Europe so that new fault lines do not open up on the continent when facing the Russian challenge, as this would only aid the Russians in the long-term competition.

"That's why I believe that one of the most important tasks of our current presidency, which Czech Republic took over the 1st of July, would be to keep Europe together and find ways to handle the crisis in Ukraine and relationships with Russia. Because if we let ourselves become divided then Russia would prevail and that was their primary objective from the very beginning. We need to bring our understanding of situation to the same foundation so that we don't take different conclusions and potentially wrong decisions."

The Czech Republic has carved out a distinctive role as Ukraine's "ammunition champion," pioneering an innovative procurement initiative that has become crucial to sustaining Ukrainian artillery capabilities. The Czech ammunition initiative emerged in early 2024 in response to Ukraine's acute shortage of artillery shells, exacerbated by delays in U.S. military aid. Czech officials identified sources for ammunition outside the traditional Western defense industrial base and coordinated an international procurement effort. According to Czech President Petr Pavel, the initiative delivered 1.5 million rounds of ammunition in 2024, including 500,000 large-caliber 155mm and 152mm shells.[16]

The initiative brings together more than 15 countries globally willing to purchase ammunition for Ukraine, with the Czech Republic acting as coordinator or intermediary._Prague itself contributed €34 million to the initiative. In 2025, the initiative has enjoyed greater success, with contributions increasing by 29% compared to 2024.[17] President Pavel indicated in May 2025 that Prague intended to deliver up to 1.8 million additional shells, though this depended on the results of parliamentary elections scheduled for October 2025.[18]

Beyond the ammunition initiative, Czech military support has been substantial. The country has supplied weapons, ammunition, and armored vehicles since the invasion's earliest days, with total aid reaching almost €286 million by mid-2025.[19] During the Czech Presidency of the Council of the EU in the second half of 2022, Ukraine was designated a priority, with discussions focusing on approving sanctions, cutting dependence on Russian fuel imports, and providing military aid via the European Peace Facility.

Nevertheless, Czech public opinion toward Ukraine has shown signs of declining support. According to a STEM survey conducted on July 16, 2025, 46% of Czech citizens reported being not very interested, or not interested at all, in events in Ukraine.[20] A Data Collect survey for Czech Television, conducted between February 10-14, 2025, found that while 77% of Czechs support humanitarian aid to Ukraine, around 60% favored a quick ceasefire, even if this meant territorial losses for Ukraine.[21]

These polling figures reflect broader European war fatigue and economic concerns that have become more pronounced as the conflict extends into its fourth year. The political vulnerability of Czech support underscores a broader challenge: even among strongly pro-Ukrainian governments, sustaining public backing for open-ended military assistance requires constant political effort and faces increasing headwinds from populist opposition parties.

THE CAUTIOUS SUPPORTERS: ROMANIA AND BULGARIA

Romania and Bulgaria represent a more ambivalent approach to supporting Ukraine, providing assistance but with significant constraints imposed by security concerns, domestic politics, and historical ambivalence toward Russia.

Romania's approach to the Ukraine war has been characterized by extreme caution. The country has provided military aid to Ukraine since the beginning of the full-scale invasion, but has done so largely non-publicly. Intelligence from open sources has identified

Romanian TAB-71 armored personnel carriers and APR-40 multiple rocket launchers on the battlefield, along with howitzers, 122mm and 152mm shells, grenade launchers, and DShK machine guns.[22]

The most significant Romanian contribution became public in September 2024, when President Klaus Iohannis signed documents for the complete transfer of a Patriot anti-missile system to Ukraine. The decision placed Romania among a select group of NATO countries providing fully functional Patriot systems to Ukraine, including the United States, Germany, and the Netherlands.[23]

Romanian Acting Prime Minister Ilie Bolojan noted in August 2025 that military aid packages to Ukraine consisted primarily of Soviet-era ammunition, materiel that Ukrainian forces could use with their existing weapon systems. Romania has also contributed to Ukraine's defense infrastructure by training Ukrainian pilots on F-16 fighter jets and facilitating grain exports through its ports, while supplying electricity and fuel to support Ukraine's energy sector.[24]

However, Romania's commitment has faced internal challenges. In October 2022, Defense Minister Vasile Dîncu gave a controversial interview stating that the Kremlin had the means to prolong the conflict indefinitely and that Ukraine's only chance for peace was to engage in negotiations with Russia, talks that should be conducted by international actors like NATO and the U.S. rather than Ukraine itself.[25]

Although President Iohannis quickly contradicted these statements, affirming that Romania's official stance was that "Ukraine will decide on its own when, what and how it will negotiate," the episode revealed divisions within the Romanian elite.[26]

The most serious threat to Romanian support emerged in November 2024, when pro-Russian ultranationalist Calin Georgescu won the first round of Romania's presidential elections. Georgescu's surprise victory raised the specter of an extended pro-Russian border with Ukraine, potentially creating a trio of allies, including Hungary and Slovakia, within the EU that opposed Europe's otherwise unanimous condemnation of Russia. However, Romania's Constitutional Court subsequently annulled the first-round election results amid concerns about Russian interference in the electoral process.[27]

As of March 2025, Romania's Supreme Council of National Defense reaffirmed its commitment to continue military support for Ukraine, with the council analyzing the situation related to Russia's war and paying particular attention to prospects for peace negotiations. Romania has also moved forward with expanding the military complex at the 57th Airbase "Mihail Kogălniceanu," which is planned to become the largest NATO base in Europe and serve as a stronghold for deterring Russia.[28]

Bulgaria has maintained the lowest profile among former Warsaw Pact EU/NATO members regarding Ukraine support. The country's contribution has been largely limited to participation in Black Sea security operations. Starting July 1, 2024, Romania, Bulgaria, and Turkey launched a joint operation to clear the Black Sea of mines, which benefits Ukraine's agricultural exports and maritime security.

Bulgaria's minimal engagement reflects several factors: historical and cultural ties to Russia through Orthodox Christianity and pan-Slavic sentiment, dependence on Russian energy (with leaked documents confirming Russia's concealed control over the TurkStream pipeline running through Bulgaria), and a powerful pro-Russian political faction domestically.[29]

These constraints have limited Bulgaria's willingness to provide substantial military or financial support to Ukraine, making it an outlier among former Warsaw Pact states within NATO and the EU.

THE OBSTRUCTIONISTS:
HUNGARY AND SLOVAKIA

The most striking feature of the regional response to the Ukraine war has been the emergence of an actively obstructionist bloc led by Hungary's Viktor Orbán and Slovakia's Robert Fico.

These leaders have not merely refrained from supporting Ukraine; they have actively worked to undermine European and NATO assistance efforts, align themselves with Russian narratives, and maintain close ties with Moscow even as it prosecutes a war of aggression against a European democracy.

Hungary, led by Viktor Orbán, is indeed the only EU and NATO country that has categorically refused to supply arms to Ukraine or allow weapons transports across its territory, and has frequently opposed the collective Western approach to supporting Kyiv against Russian aggression. Orbán's government has repeatedly blocked or resisted joint EU and NATO measures, justified by a combination of calls for peace, demands for negotiated settlements, and claims that Western strategy is failing.

Orbán regularly argues that the Western military and financial support to Ukraine "is not working." In October 2023, he stated, "the Ukrainians will not win on the front line," and called for a new strategy or "plan B," a position interpreted as advocating for a compromise peace that may require Ukraine to make territorial concessions. These remarks were widely reported as diverging sharply from the mainstream EU and U.S. stance, and further reinforced Hungary's outlier position.

By October 2025, Orbán's position hardened, including during and after his participation at the EU summit in Copenhagen. He publicly declared, "We do not want to die for Ukraine," and initiated a Hungarian signature campaign in opposition to EU war policy, arguing that "the risk of being dragged further into the war grows" as the EU continues its support for Ukraine. This campaign has been reported in Hungarian and international media as part of a broader effort to build domestic and regional opposition to Brussels's actions.[30]

Hungary's veto power within the EU has enabled Orbán to block or delay major aid packages to Ukraine. Notably, in December 2023, Hungary was the only member state to veto the €54 billion multi-year Ukraine aid package, insisting the aid should be disbursed in annual tranches subject to yearly reviews. Moreover, Orbán has voiced skepticism of Ukraine's path to full EU membership, instead proposing alternative frameworks similar to those used in the EU's relationships with the UK or Turkey.

At a special EU summit in March 2025, Orbán blocked joint conclusions supporting Ukraine, forcing the other 26 leaders to bypass Hungarian endorsement. European Council President António Costa commented, "Hungary is isolated... 26 are more than one".[31]

Hungarian rejection also extends to Ukraine's NATO membership ambitions; in January 2025, Orbán and Slovak PM Fico publicly pledged to oppose Ukraine's NATO entry, with Orbán asserting such accession "never will be" on the agenda.[32]

Slovakia's trajectory has been particularly dramatic, shifting from significant supporter to active opponent of Ukraine aid following Robert Fico's election as prime minister in October 2023. Prior to Fico's ascent, Slovakia had been a substantial supplier of military equipment to Ukraine, providing €671 million worth of military aid between February 2022 and October 2023.[33] Slovakia operated a repair facility for damaged Ukrainian military equipment in the city of Michalovce and had transferred significant quantities of weapons, including combat vehicles, air defense systems, and MiG-29 fighter jets.[34]

This support ceased abruptly after Fico's election. During his 2023 campaign, Fico pledged "not to send another bullet to Ukraine" and later claimed provocatively that "there's no war in Kyiv."[35] By October 2024, Fico explicitly ruled out further military aid to Ukraine and additional sanctions on Russia. His position hardened further when he accepted Putin's invitation to attend the 2025 Victory Day celebrations in Moscow, an event commemorating the Soviet Union's triumph over Nazi Germany that Putin has explicitly linked to his war against Ukraine.[36]

Fico has echoed Russian narratives about the conflict, arguing that additional military support for Kyiv "only prolongs the conflict" and that the West is "falsely demonizing Putin." In a January 2024 op-ed, Fico claimed his EU and NATO allies "made a huge mistake" in dismissing early calls for a ceasefire on Russian terms, arguing that Western powers "incorrectly evaluated the use of Russian military force as an opportunity to bring Russia to its knees."[37]

Like Orbán, Fico has used Slovakia's position within the EU to obstruct collective support for Ukraine. In March 2025, Fico threatened to block EU statements during a key summit unless the bloc called for an immediate ceasefire and launched talks with Russia.[38] He stated: "If the summit does not respect that there are other options besides simply continuing the war, the European Council may not be able to agree on conclusions regarding Ukraine on Thursday." Though Fico was ultimately won over after some modifications to the text, his willingness to obstruct European unity has become a consistent pattern.

At his January 2025 meeting with Orbán in Bratislava, Fico pledged that Slovakia would "never approve Ukraine's membership in NATO," though he stated Slovakia supports Ukraine's potential EU membership "provided the necessary conditions are met."[39]

The alignment between Hungary and Slovakia has created a pro-Russian axis within the EU and NATO, a development with potentially serious consequences for European unity. The two countries have coordinated their positions on blocking or delaying aid to Ukraine, opposing Ukraine's NATO membership, and maintaining energy ties with Russia.

This coordination represents a fundamental challenge to the consensus-based decision-making processes of both the EU and NATO, as either country can exercise veto power to obstruct collective action.

And the admiration which American conservatives have for Orbán sends an impression that this extends to his policies on Ukraine and thus suggesting that the Trump Administration is ambivalent about Ukraine's success.

This impression has of course been generated by Trump himself, but in the presence of Putin's unwillingness to even accept an idea of the viability of Ukraine is leading to Trump seeking a path to end the war but clearly not on Putin's terms.

The divisions over Ukraine have effectively destroyed the Visegrad Group (V4) which was the informal alliance among Poland, the Czech Republic, Hungary, and Slovakia that had been a significant force within EU and NATO since its founding in 1991.

The summit of the four prime ministers in Prague on February 27, 2024, laid bare the fundamental divisions. On one hand, the Czech Republic and Poland argued for even more and faster arms deliveries to Ukraine. On the other, Hungary and Slovakia refused to send weapons and wanted "peace" at seemingly any cost, with Hungary having developed a distinct policy of watering down Russia sanctions, questioning Ukraine's EU integration, and at one point blocking EU aid for Kyiv.[40]

CONCLUSION

As of October 2025, the former Warsaw Pact members of the EU and NATO are divided into three distinct camps with fundamentally different approaches to the Ukraine war:

- The Frontline Maximalists (Poland, Estonia, Latvia, Lithuania): These countries provide substantial military and financial aid relative to their size, advocate for more aggressive support including potential troop deployments, view Ukraine's victory as essential to their own security, and maintain the highest defense spending as a percentage of GDP within NATO. Their position is rooted in geographic proximity to Russia, historical experience of Soviet occupation, and clear-eyed assessment that Russian victory in Ukraine would likely lead to further aggression.
- The Cautious Supporters (Czech Republic, Romania): These countries provide significant but politically vulnerable support, face domestic opposition or war fatigue, maintain some public reservations about aid levels or approach, and balance support for Ukraine with concerns about escalation and their own defense needs. Their support, while substantial, remains contingent on political developments and public opinion.

- The Obstructionist Bloc (Hungary, Slovakia), with Bulgaria as a minimal participant: These countries actively oppose or obstruct military aid to Ukraine, maintain or advocate for maintaining ties with Russia, use veto power to block or delay EU/NATO initiatives, promote narratives about negotiated settlement that would likely require Ukrainian territorial concessions, and oppose Ukraine's NATO membership while showing ambivalence about EU membership.

This tripartite division has profound implications for European security and the transatlantic alliance. The frontline states have warned repeatedly that Russian success in Ukraine would embolden Moscow to test NATO's Article 5 commitments, likely in the Baltic region.

The European Union and its member states have collectively provided over $186 billion in financial, military, and humanitarian aid to Ukraine since the Russian full-scale invasion, with military assistance amounting to at least $51 billion according to key sources tracking these commitments.[41] This support has not been without controversy: Hungary and Slovakia have repeatedly used their veto powers to block, delay, or weaken EU measures to support Ukraine, exploiting the requirement for unanimity in critical decisions.

These internal divisions have resulted in significant delays to military and financial assistance packages, as documented in multiple high-level statements and news coverage from May to October 2025. As a consequence, the obstructionist bloc's actions have had measurable battlefield impacts due to postponed deliveries of weapons, funds, and other crucial supplies for Ukraine's defense.

The disproportionate influence of Hungary and Slovakia on EU policy toward Ukraine is a direct function of the Union's unanimity rule, giving them the ability to shape or obstruct collective action, despite representing a small minority of member states

The response of former Warsaw Pact EU and NATO members to the Ukraine war underscores that the Soviet legacy, rather than creating a unified response to Russian aggression, has produced divergent reactions based on geography, domestic politics, energy dependencies, and competing narratives about European identity and sovereignty.

The frontline states, Poland and the Baltic countries, have demonstrated that some nations have genuinely internalized the lessons of the Cold War and Soviet domination, translating historical experience into contemporary security policy. Their outsized contributions to Ukraine's defense reflect not altruism but strategic necessity: they understand that their own security depends on Ukraine's success in repelling Russian aggression.

The obstructionist bloc led by Hungary and Slovakia represents a different interpretation of history and national interest, one that prioritizes short-term economic considerations, domestic political advantage, and a transactional relationship with Russia over solidarity with a fellow European democracy under attack. This position has found resonance with certain segments of their populations but has also isolated these countries within European institutions.

The cautious supporters occupy an uncomfortable middle ground, providing assistance while hedging their bets and facing growing domestic skepticism about the sustainability and wisdom of open-ended support.

The former Warsaw Pact states, having escaped Soviet domination three decades ago, now face a fundamental question: will they stand together against Russian imperialism or will old dependencies and new divisions allow Moscow to rebuild its sphere of influence through military aggression and political manipulation?

And if there is an armistice in Ukraine, can these states provide leadership for how to manage the future of Europe with both Ukraine and Russia?

SECTION V
ASIAN DYNAMICS

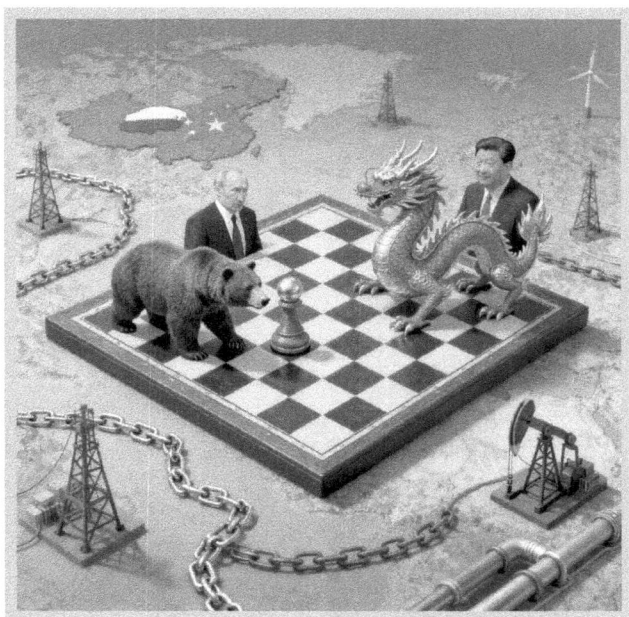

CHAPTER 1
CHINA'S STRATEGIC SUBORDINATION OF RUSSIA

In the complex chess game of global geopolitics, few developments have been as paradoxical as the way Vladimir Putin's war in Ukraine has inadvertently strengthened China's strategic position while simultaneously transforming Russia into Beijing's economic subordinate.

What was once promoted as a "no limits" partnership between equals has become an increasingly asymmetrical relationship where China holds decisive advantages, capitalizing on Russia's isolation to establish economic and strategic hegemony over its northern neighbor.

By 2022, Xi Jinping confronted a daunting economic landscape. China's growth model, built on manufacturing exports and massive infrastructure investment, was showing signs of strain. The property sector was in crisis, demographic trends pointed toward a shrinking workforce, and Western companies were increasingly "friend-shoring" their production to India, Mexico, or back to domestic facilities.

The Chinese government's emphasis on maintaining political control through tech crackdowns and regulatory interventions often worked against the market dynamism needed to drive consumption and innovation. Putin's invasion of Ukraine in February 2022 unexpectedly provided Xi with a pathway through these challenges.

While Western sanctions isolated Russia economically, China positioned itself as Moscow's indispensable economic lifeline. Through systematic exploitation of Russia's isolation, Beijing has transformed a proclaimed partnership into a relationship of dependence spanning currency, energy, and market access.

THE YUAN'S RISE: FROM MINOR PLAYER TO ECONOMIC LIFELINE

The most striking manifestation of Russia's growing dependence lies in the rapid "yuanization" of the Russian economy. Before the Ukraine invasion, the Chinese yuan accounted for less than 2% of Russia's international trade settlements. In 2025, it has become what Russian Central Bank Governor Elvira Nabiullina describes as the "primary" foreign currency for Russia's overseas economic activity.[1]

The transformation has been dramatic. On the Moscow Stock Exchange, yuan transactions surged from a mere 3% in 2022 to 54% by May 2024. Following U.S. sanctions on the Moscow Exchange in summer 2024, yuan transactions reached an extraordinary 99.8% of all foreign currency trading. By December 2024, nearly 90% of all Russia-China transactions were reportedly settled in yuan and rubles.[2]

This shift represents far more than simple currency substitution. Russia has essentially swapped its dollar dependence for reliance on the yuan, meaning Russian reserves and payments are now influenced by the policies of the Chinese Communist Party and the People's Bank of China. The implications are profound: should relations deteriorate, Russia faces potential reserve losses and payment disruptions at Beijing's discretion.[3]

The vulnerability became apparent when Chinese banks, concerned about secondary U.S. sanctions, began restricting yuan transactions with Russian counterparts in 2024. Major Chinese financial institutions, including ICBC, Bank of China, and China CITIC

Bank, ceased processing yuan-denominated transactions from Russia.

This left Russian companies scrambling for yuan liquidity, with some transactions stuck in limbo for months. The Russian Central Bank's admission that currency swaps are meant only for "short-term stabilization" rather than long-term funding underscored Moscow's precarious position.[4]

ENERGY RELATIONS: THE ILLUSION OF PARTNERSHIP

While Russia often portrays its energy relationship with China as a successful pivot away from Western markets, the reality reveals a more subordinated position. Chinese imports of Russian energy soared following the invasion, often at heavily discounted prices, providing Beijing with crucial raw materials at favorable terms during a period of global inflation. China became Russia's largest buyer of fossil fuels, accounting for 42% of Russia's monthly export earnings from its top five importers. However, this dominance comes at a significant cost to Russian leverage.[5]

Beijing has systematically extracted substantial discounts on Russian energy exports. According to Bank of France analysis, Chinese refineries have benefited from discounts on Russian oil ranging from 10% to 19% compared to other suppliers. These discounts started in April 2022 and peaked in July of that year at 19% below market rates. From January to June 2023, Chinese imports of Russian oil remained consistently below Brent prices, averaging a 12.7% discount compared to other suppliers.[6]

The price advantage extends beyond crude oil. China purchased 44% of Russia's coal exports from December 2022 through July 2025, and 47% of Russia's crude exports over the same period. Russia's need for alternative markets following European sanctions has essentially made it a price-taker rather than a price-setter in its most crucial export relationships.

China's approach reflects calculated opportunism rather than partnership solidarity. Beijing supplies not what Russia needs, but what it can make money on. The value of dual-use civilian and military goods sold to Russia by China increased by only $1 billion in 2023, despite total exports rising 47% to $111 billion. This suggests China prioritizes profitable commercial relationships over supporting Russia's military capabilities.[7]

EXPANDING INFLUENCE BEYOND RUSSIA

More strategically significant than the direct economic benefits has been how the war accelerated Russia's decline and weakened Moscow's influence over its traditional sphere. In Central Asia, countries that had long balanced between Russian and Chinese influence found themselves dealing with an increasingly weakened northern neighbor. Kazakhstan, Uzbekistan, and other Central Asian states began tilting more decisively toward Chinese investment and trade partnerships.

As James Durso noted in a 2023 article:

Xi unveiled his vision for Central Asia that was, like most top-level political documents, laden with vague declarations like "helping each other," "common development," "universal security," "shared future," and that old Beijing favorite, "win-win."

China's strategy for Central Asia is to secure economic gains and help alleviate the high youth unemployment rate, create prosperity in the Xinjiang region to calm separatist tensions, and to build an alternate trade corridor in the event of a U.S.-led naval blockade of China's maritime trade routes.

Kazakhstan's president, Kassym-Jomart Tokayev, responded "We consistently assert that Central Asia is a creation place. We oppose turning the region into a place of geopolitical confrontation" and Turkmenistan's president, Serdar Berdimuhamedov, added, "As emphasized, the peoples of our countries have centuries-old friendly relations, a vast experience of interaction and good neighborliness."

But despite all the official bonhomie – and group pictures – the leaders put pen to paper and, according to Silk Road Briefing, "...approved US\$3.72 billion in regional grants, signed 54 major multilateral agreements, created 19 new regional platforms and signed a further 9 multilateral cooperation documents."

Among the agreed points were to coordinate China's Belt and Road Initiative with the republics' national development strategies, upgrade border checkpoints, expand agricultural exports to China, award scholarships to Central Asian students to study in China, and develop cooperation in the fields of irrigation and green energy.

Kazakhstan and China signed 47 agreements worth \$22 billion which should increase trade, a record \$31 billion in 2022, and investment. China is one of the five biggest investors in Kazakhstan with a total investment exceeding \$23 billion. Kazakhstan's exports to Russia increased 15% in 2022, and trade turnover by September 2022 was \$18.4 billion, but Astana would be more comfortable if Moscow adopted the philosophy "When you're only No. 2, you try harder" in its dealings with its now-wary Central Asian neighbor.

Uzbekistan and China adopted a "comprehensive strategic partnership" and inked 41 official documents, and Chinese and Uzbek businesses concluded \$25 billion in deals. The official agreements addressed joint efforts in higher education, alternative energy, agricultural innovation, hydroelectric power, and logistics.

A "comprehensive strategic partnership" is just below a "comprehensive strategic co-operative partnership," which is generally regarded as the highest level of bilateral relations for China. Uzbekistan now joins Kazakhstan and Turkmenistan as the third Central Asian republic at this level of engagement with China, giving China privileged access to the two biggest economies in Central Asia, and the country with 10% of the world's natural gas reserves, that all sit astride Eurasia's East-West transport links.[8]

This shift allowed China to expand its Belt and Road Initiative influence in a region rich with energy resources and strategically positioned between China and European markets. What emerged was a fundamental reshaping of Eurasian geopolitics in China's favor, an outcome that might have taken decades to achieve through purely economic competition with a stronger Russia.

The territorial question adds another layer to the subordination dynamic. In August 2023, China's Ministry of Natural Resources published a map showing Bolshoy Ussuriysky Island entirely as Chinese territory, despite a 2004 agreement dividing the island roughly in half. This cartographic assertion reflects deeper historical grievances about 19th-century "unequal treaties" that transferred approximately 6 million square kilometers of territory to Russia.[9]

Beijing has become the largest trading partner and foreign investor in the Russian Far East, pouring significant funds into infrastructure, energy, agriculture, and port development.[10] This economic penetration, combined with demographic concerns about Chinese migration to the region, creates long-term leverage that transcends immediate territorial disputes.

THE ASYMMETRICAL PARTNERSHIP

For Russia, China represents economic salvation from Western sanctions, providing alternative markets for energy exports and essential imports. The relationship has become existential, as without China's assistance, Russia would struggle to finance its war effort or secure necessary resources for military operations. The yuan has become so central to Russian economic activity that Moscow's financial system now operates essentially within a Chinese currency zone.

For China, however, Russia remains a relatively minor economic partner whose primary value lies in providing discounted energy and serving as a testing ground for yuan internationalization. Chinese trade with Russia, while reaching record levels, represents a small fraction of China's global economic activity. This asymmetry grants Beijing tremendous leverage to extract favorable terms while limiting its own commitments.

China's economic subordination of Russia demonstrates considerable agility in adapting to changing global circumstances, turning potential complications into opportunities for expanded influence.

By maintaining carefully calibrated support for Moscow while extracting maximum economic benefit, Beijing has positioned itself as Russia's indispensable partner while avoiding the costs of genuine alliance.

For the West, China's subordination of Russia consolidates more power in Beijing's hands and provides China with valuable resources at favorable prices. The relationship serves as a proof of concept for Chinese economic statecraft, demonstrating how Beijing can leverage economic dependencies to achieve strategic objectives.

However, this narrative of Chinese strategic gains should be tempered with recognition of its limitations. China's closer relationship with Russia, while economically beneficial in the short term, carries significant risks.

Secondary sanctions remain a constant threat, and the association with an increasingly isolated regime complicates China's relationships with European and other Western partners. The cost of reduced access to Western technology and investment may outweigh the benefits of expanded trade relationships elsewhere.

CHAPTER 2
PUTIN'S BEIJING VISIT: SEPTEMBER 2025

In a display of geopolitical theater that captured global attention, Russian President Vladimir Putin stood shoulder-to-shoulder with Chinese leader Xi Jinping and North Korean dictator Kim Jong Un at Beijing's Tiananmen Square on September 3, 2025.

The occasion was China's massive military parade commemorating the 80th anniversary of Japan's surrender in World War II, but the real message was unmistakably contemporary: the emergence of a powerful Eastern bloc challenging Western dominance.

This marked the first time the three leaders had appeared together in public, creating what many analysts describe as a visual symbol of an "Axis of Upheaval," that is countries united not necessarily by ideology, but by their shared opposition to Western-led global order. The carefully choreographed event drew more than 50,000 spectators to Tiananmen Square and was viewed by 1.9 billion people online and over 400 million on television, according to Chinese state broadcaster CCTV.

Putin's visit to Beijing was far from a ceremonial appearance. Spanning four intensive days, the Russian leader engaged in what he described as a "multi-day visit indeed" that encompassed several critical events: the Shanghai Cooperation Organization (SCO) summit, a trilateral Russia-Mongolia-China meeting, and bilateral discussions with Xi Jinping. This packed schedule was deliberately designed to maximize diplomatic efficiency while avoiding multiple long-distance trips.

The format proved highly effective for Putin's objectives. As he explained in his concluding press conference, this arrangement was "not only good for meeting at the negotiating table but, more importantly, for holding many informal discussions on any issue of mutual interest in an informal and friendly atmosphere." These informal exchanges often prove more valuable than formal negotiations in advancing strategic partnerships.

One of the most significant outcomes of Putin's visit was the signing of memorandum of understandings related to the Power of Siberia 2 gas pipeline project. Putin emphasized that this project had been years in the making, with negotiations examining multiple routes and weighing various pros and cons. The agreement reflects the growing energy demands of the Asia-Pacific region, particularly China's economy, which continues to post over five percent growth. "The global economy is nevertheless growing, especially in the Asia-Pacific region," Putin noted, highlighting the economic logic driving this partnership.[1]

Through Mongolia, Russia will supply 50 billion cubic meters of gas, adding to the existing 38 billion cubic meters, with additional routes planned to bring the total to over 100 billion cubic meters. For China, this arrangement provides steady, reliable energy supplies at balanced market prices, creating significant competitive advantages over European markets that face higher, less stable pricing.[2]

Nonetheless, the actual pricing structure has not been finalized. Reports indicate that pricing details remain unresolved and are still under negotiation between Russia and China. There are significant disagreements over pricing, with Russia wanting prices linked to Asian oil benchmarks (around \$265-285 per 1,000 cubic meters) while China seeks much lower prices closer to Russia's domestic rates (around \$120-130).[3]

The military parade itself served as a powerful demonstration of China's advancing military capabilities and the solidarity among leaders challenging Western influence. The spectacle featured advanced weaponry including intercontinental nuclear missiles, anti-ship missiles, undersea drones, air-defense lasers, and even robot dogs, showcasing what Chinese officials described as equipment "at the forefront of global standards."

Xi Jinping's recent speech at the 80th anniversary military parade in Beijing underscored stark global choices, declaring that humanity faces a crossroads between "peace or war" and "dialogue or confrontation", words interpreted internationally as a veiled warning referencing Taiwan and broader geopolitical tensions. Without naming Taiwan directly, Xi emphasized the role of the Chinese military in safeguarding China's sovereignty and territorial integrity, reiterating Beijing's well-documented claim over the self-governing island and its readiness to resist external interference.[4]

Xi stressed the importance of learning from the past to avert future conflicts, arguing that the Chinese people stand on "the right side of history" and remain committed to peaceful development. He projected confidence in China's trajectory and positioned Beijing as a stabilizing force preferred over confrontation or zero-sum thinking in international relations.

Although Taiwan was not named, Xi's remarks on territorial integrity were widely interpreted as a message to Taipei, corresponding with Beijing's ongoing military modernization and warnings against "Taiwan independence" and "external interference," especially by Western governments. Xi and other Chinese officials consistently frame Taiwan as a "sacred" part of China and call for reunification, by peaceful means if possible, but with force not ruled out if necessary. Recent military exercises near Taiwan, coupled with these strong rhetorical signals, illustrate a coordinated pressure campaign amid rising cross-strait tensions and ongoing resistance from Taipei's leadership.[5]

Xi's speech fits within the broader Chinese strategy of asserting sovereignty and countering perceived external threats to national interests. This narrative resonates with Beijing's declared intentions to modernize its military by 2035 and achieve "world-class" status by 2049, with clear references to deterrence, readiness, and long-term planning around the Taiwan question.[6]

The presence of foreign leaders at the parade, and parallel statements from Taiwan condemning China's display of power, highlight the contested narratives surrounding sovereignty and peace in the region. Taiwan continues to assert its right to self-determination and rejects Beijing's linkage of peace with military might.

In other words, Xi Jinping's parade speech deploys stark rhetoric on the global choice between peace and confrontation, uses indirect language to reinforce demands over Taiwan, and places China's military front-and-center in defending these interests, all of which point to increasing assertiveness in the cross-strait and international domains.

Putin's visit occurred against the backdrop of escalating Western sanctions and economic pressure. During his press conference, he addressed European preparations for a 19th package of sanctions, this time targeting not just Russia but its partners. His response revealed a strategic calculation: that sanctions are being used not primarily to address the Ukraine conflict, but to resolve broader economic competitions with countries whose "economic ties and advantages do not suit someone."

Putin pointed to examples beyond Russia, noting trade imbalances between the United States and India or China, and even mentioning sanctions imposed on Brazil despite the lack of any apparent connection to Ukraine. This framing positions Western sanctions as economic warfare designed to maintain American advantage rather than principled responses to specific conflicts.

The resistance strategy emerging from this Eastern alliance relies on the economic heft of its members. Putin noted that countries like India and China, with populations of 1.5 billion and 1.3 billion respectively, "boast powerful economies and live by their own domestic political laws." These nations have historical experience with colonialism and attacks on their sovereignty, making them resistant to external pressure tactics.

Throughout his Beijing visit, Putin addressed questions about the ongoing conflict in Ukraine and prospects for peace negotiations. His responses revealed both flexibility and firm red lines. When asked about meeting with Ukrainian President Volodymyr Zelensky, Putin indicated willingness but questioned the legal legitimacy of such discussions given constitutional issues surrounding Zelensky's extended presidency under martial law.

Putin's constitutional argument is detailed and specific: Ukrainian law provides no mechanism for extending presidential powers beyond the five-year term, and while elections cannot be held under martial law, this does not automatically extend the president's authority. He argued that according to Ukrainian constitutional requirements, power should transfer to the speaker of the Rada, including supreme commander-in-chief responsibilities.[7]

Despite these legal complications, Putin expressed openness to properly prepared meetings that could lead to positive results. He revealed that President Trump had specifically asked him to consider meeting with Zelensky, to which Putin responded positively: "if Zelensky is ready, he can come to Moscow, and we will have such a meeting."

Of course, one could not be certain that such a visit would prove fatal for Zelensky.

On the broader conflict, Putin maintained that Russia's objectives remain unchanged from the beginning of the special military operation. He described Russian forces as "advancing in all directions" at different paces, while Ukrainian forces are increasingly running out of reserves, with combat-ready units staffed at only 47-48 percent capacity.

The Russian leader revealed that during his four days in Beijing, "all of my dialogue partners without exception" were supportive of the Alaska meeting between Putin and Trump, and "all of them expressed hope that the position of President Trump and the position of Russia and other participants in the negotiations will put an end to the armed conflict."[8]

The gathering in Beijing prompted sharp responses from regional democracies and Western allies. Taiwanese President Lai Ching-te used the occasion to warn about fascism under Xi's rule, noting that "Taiwan does not use guns to commemorate peace" and defining fascism through "extreme nationalism, repression of freedom of speech, and obvious strongman leader worship."[9]

European leaders expressed alarm at the visual symbolism of the three leaders standing together. German Chancellor Friedrich Merz went so far as to call Putin "perhaps the most serious war criminal of our time," a comment Putin dismissed as "an unsuccessful attempt to shift responsibility" for the Ukraine crisis from Western actions that, in his view, precipitated the conflict.[10]

The broader European concern reflects recognition that this gathering represents more than ceremonial diplomacy. It signals the emergence of an alternative power structure that could fundamentally alter global governance. The economic dimensions of this emerging alliance extend far beyond energy cooperation.

Putin repeatedly asserts that trade problems should be solved "in a negotiation process," praising long consultations and joint efforts to overcome obstacles as yielding solutions over time. In public forums, he stresses that Russia endeavors to understand what each partner can offer and looks for ways to increase imports of those items, reflecting a pragmatic, flexible style instead of confrontational trade policy.

This approach allows Russia to maintain stable relationships with key markets, support the emergence of new partnerships, and prevent unnecessary retaliation or trade conflict. It also contributes to the "global economic rebalancing" Putin often refers to, as Russia and partners adapt nimbly to evolving supply and demand patterns, sanctions, or geopolitical influences rather than defaulting to protectionist measures.

In other words, Putin's discussion of trade relationships underscores a deliberate preference for constructive negotiation and gradual adjustment, even when faced with large trade imbalances, instead of resorting to threats or sharp policy shifts.[11]

Putin's four-day visit to Beijing achieved its immediate objective of demonstrating that Russia is not internationally isolated, but at a cost that reveals the fundamental weakness of Russia's current position. While the visual symbolism of standing alongside Xi Jinping and Kim Jong Un provided propaganda value for domestic consumption, the substance of the agreements exposed Russia's relegation to junior partner status in what Putin characterized as an equal partnership.

The pipeline agreements that Putin presented as major achievements are, in reality, arrangements that burden Russia with massive infrastructure investments while providing China with decades of discounted energy. The broader economic relationship sees China extracting maximum value from Russian resources and markets while carefully avoiding commitments that would trigger Western economic retaliation.

Xi Jinping's balancing act allows China to benefit from Russian desperation while preserving the economic relationships with Europe and the West that remain crucial to Chinese growth. This approach demonstrates that while Putin may speak of multipolarity and equal partnerships, the reality is a relationship where China increasingly calls the shots and Russia provides the resources.

The broader implications suggest that rather than creating a genuine alternative to Western-led institutions, Putin's diplomatic efforts have primarily succeeded in making Russia more dependent on Chinese goodwill. The "strategic partnership" rhetoric obscures what is essentially a relationship of dependency, where Russia's international isolation has created opportunities for China to advance its own interests while providing just enough support to keep Russia engaged.

In short, Putin's Beijing visit represents not the emergence of a balanced multipolar order, but rather the crystallization of a new hierarchy where China assumes an increasingly dominant position over a weakened and isolated Russia. Putin's Russia finds itself in the unfamiliar position of being the supplicant rather than the power broker.

CHAPTER 3
ASIAN POWERS AND
THE UKRAINE CONFLICT

T he war in Ukraine has fundamentally transformed global security dynamics in ways that extend far beyond European borders. While much attention has focused on Western support for Ukraine and Russian aggression, the conflict has catalyzed significant shifts across Asia, bringing three key players, Japan, South Korea, and North Korea, into unprecedented roles on the world stage.

Their involvement reveals how Putin's invasion has inadvertently accelerated geopolitical realignments, technological developments, and the emergence of new international partnerships that will shape the 21st century security landscape.

JAPAN'S HISTORIC
INTELLIGENCE PARTNERSHIP

Perhaps no development better illustrates the far-reaching consequences of the Ukraine war than Japan's decision to share satellite intelligence with Ukraine, a move that represents a watershed moment in Tokyo's post-war foreign policy.

For former Japanese Prime Minister Kishida, the security of

Europe and the Indo-Pacific region are "inseparables," considering that "today's Ukraine could be tomorrow's East Asia."

Several factors explain Japan's evolution:

- Rising tensions in the Indo-Pacific, particularly since the July 2024 strategic partnership treaty between North Korea and Russia.
- Growing integration into the Quad alliance with the United States, Australia, and India.
- And recognition that the Ukrainian conflict has demonstrated the interconnection of European and Asian security theaters.

On April 21, 2025, Japan announced its first geospatial intelligence sharing agreement with Ukraine, marking a dramatic departure from decades of strict limitations on defense exports and intelligence cooperation.[1]

The partnership centers on synthetic aperture radar (SAR) imagery from the Institute for Q-shu Pioneers of Space (iQPS), an innovative Japanese company that has revolutionized satellite intelligence through dramatic cost reduction and technical innovation. The agreement provides Ukraine with high-definition images of 46 centimeters resolution usable day and night in all weather conditions from a constellation projected to expand from five satellites to 24 by 2027, with an ultimate goal of 36 satellites by the 2030s.[2]

The catalyst for this unprecedented cooperation arrived in February 2025, when the United States temporarily suspended intelligence sharing with Ukraine. Ukrainian officials, suddenly finding themselves partially blind on the digital battlefield, reached out to allies across Europe and discovered that no single European operator could match the quantity and quality of intelligence traditionally supplied by American satellites. Germany's SAR-Lupe and SARah systems, Finland's ICEYE constellation, and Italy's COSMO-SkyMed satellites were providing imagery, but coverage remained incomplete.

Japan's SAR technology offered something unique: unlike optical satellites requiring clear skies and daylight, SAR satellites penetrate clouds, darkness, and dense foliage by bouncing microwave pulses off Earth's surface. For Ukraine's Main Intelligence Directorate, this capability proved invaluable for monitoring Russian positions during harsh winters, tracking supply convoys through forest cover, and assessing damage through smoke and debris.[3]

What makes iQPS particularly remarkable is its economic revolution in space intelligence. The company achieved a technical breakthrough by developing SAR satellites weighing only 100 kilograms and costing one-hundredth the price of conventional platforms. This stems from a patented 3.6-meter deployable antenna that folds compactly at launch, then deploys in space using spring mechanisms. This bowl-shaped antenna maintains curvature precision that minimizes signal distortion, enabling the high-resolution imagery.[4]

This partnership fits within a broader framework established by the Japan-Ukraine bilateral security agreement of June 13, 2024, which explicitly provides for "cooperation in the fields of intelligence and protection of classified information," and the information security agreement signed November 16, 2024, which entered into force on June 21, 2025. But the actual implementation of satellite intelligence sharing represents the true historic turning point.[5]

SOUTH KOREA'S DEFENSE EXPORTS AS DIPLOMATIC POWER

While Japan's involvement centers on intelligence and technology sharing, South Korea has taken a different path in supporting Ukraine, one that exemplifies its transformation from a middle power dependent on foreign military aid to what former President Yoon Suk-yeol termed a "Global Pivotal State."[6] South Korea's role has been shaped primarily through an unprecedented boom in defense exports that has fundamentally reshaped its international relationships and strategic capabilities.

The scale of growth is remarkable: from modest annual exports of $2-3 billion until 2020, the industry exploded to $7.3 billion in 2021, peaked at $17.3 billion in 2022, and maintained $14 billion in 2023. This trajectory propelled South Korea to become the world's 10th largest arms exporter, capturing 2% of the global arms market, a 12% increase over the past decade.[7]

The timing coincided precisely with the Ukraine war outbreak, as European nations rushed to modernize military capabilities and replenish weapons stocks being transferred to Ukraine. South Korea's defense industry found itself uniquely positioned to meet this demand through what analysts describe as a "cheaper, better, faster" value proposition compared to traditional Western suppliers.[8]

Notably, the $22 billion framework agreement signed with Poland in 2022, immediately following Russia's invasion, represents the most significant manifestation of South Korea's new role in European security.[9] This deal includes K2 Black Panther main battle tanks, K9 Thunder self-propelled howitzers, Chunmoo multiple rocket launchers, and FA-50 light combat aircraft. Beyond simple arms sales, the agreement includes technology transfers, local production arrangements, and civilian infrastructure components.[10]

Poland's strategic importance extends beyond immediate commercial value. As a frontline NATO state bordering Ukraine, Poland has become a critical logistics hub for Western military aid to Ukrainian forces. South Korean weapons flowing to Poland contribute directly to efforts to strengthen NATO's eastern flank and maintain the flow of military support to Ukraine.

South Korea's defense exports carry crucial strategic advantages for NATO countries. Due to regular joint military exercises with the United States, Korean weapon platforms maintain high compatibility with U.S. and NATO systems, allowing countries to simplify logistics when acquiring Korean weapons.

This interoperability has proven especially valuable for NATO countries seeking to rapidly modernize forces in response to the Ukraine conflict. The compatibility factor extends beyond technical specifications to operational doctrine and training, as South Korean military systems integrate seamlessly with existing NATO command and control structures.

This arms export growth provides South Korea with new tools for diplomacy and international influence. Defense exports serve as tangible demonstrations of South Korea's commitment to liberal democratic values while providing economic leverage and enhanced diplomatic relationships with key allies facing security challenges from authoritarian powers.

NORTH KOREA'S DIRECT MILITARY ENGAGEMENT

In stark contrast to South Korea's indirect support through arms exports and Japan's intelligence sharing, North Korea has chosen direct military involvement in the Ukraine conflict, marking perhaps the most significant shift in North Korean foreign policy since the Korean War and demonstrating a willingness to project military power far beyond the Korean Peninsula.

North Korea's involvement began with arms transfers to Russia in September 2023, evolving from primarily transactional relationships to strategic military partnership. According to South Korean intelligence estimates, North Korea supplied Russia with approximately 20,000 containers of weaponry by October 2024, primarily artillery shells, short-range missiles, and anti-tank missiles.[11]

The relationship was formalized through a "comprehensive strategic partnership" treaty signed in June 2024 during Russian President Vladimir Putin's first visit to North Korea in 24 years. This agreement included mutual defense provisions and marked a qualitative change in the Moscow-Pyongyang relationship.[12]

The North Korea-Russia partnership extends beyond immediate military cooperation to encompass broader strategic alignment. Russia's veto of UN resolutions monitoring North Korean sanctions violations, combined with China's abstention, has effectively weakened the international sanctions regime against North Korea.

This demonstrates how major power competition has undermined previously united international efforts to constrain North Korea's weapons programs. The partnership provides North Korea with access to advanced Russian military technologies, including space vehicles, aircraft, anti-air missile systems, and unmanned aerial vehicles, accelerating North Korea's military modernization efforts across multiple domains.[13]

THREE ASIAN POWERS, THREE APPROACHES

The involvement of Japan, South Korea, and North Korea in the Ukraine conflict reveals both the global nature of modern security challenges and the divergent approaches nations take in pursuing their strategic interests.

All three countries have leveraged the Ukraine conflict to advance their international positioning, but through fundamentally different mechanisms that reflect their contrasting values and governance systems.

- South Korea's path through defense exports and alliance partnerships demonstrates how middle powers can leverage economic capabilities to enhance their international influence within the existing international order.
- Japan's intelligence sharing represents a more cautious but still significant step toward active security cooperation beyond traditional constraints.
- North Korea's direct military involvement represents an attempt to reshape that order through alignment with revisionist powers.

The Ukraine conflict has accelerated security competition in Northeast Asia by highlighting the interconnected nature of global security challenges.

- North Korea's direct involvement in European security affairs demonstrates that regional conflicts can no

longer be contained within traditional geographic
boundaries.

- South Korea's response has been to strengthen
  partnerships with like-minded democracies globally,
  moving beyond its traditional focus on peninsula affairs.
- Japan's shift reflects recognition that its security is
  intimately connected to the broader global struggle
  between democratic and authoritarian governance
  models.

For Japan, the spillover impact from the war in Ukraine to the
Indo-Pacific is significant. The country's participation represents a
new model for international security cooperation, one where technological innovation and private sector capabilities could reshape the
balance of power in conflicts around the world.

South Korea's enhanced defense partnerships represent a significant evolution in alliance structures beyond traditional bilateral relationships. The country's defense exports have created new
multilateral security networks that extend far beyond the US-ROK
alliance. Countries receiving South Korean weapons systems become
part of an extended security network that enhances interoperability
and shared strategic interests. This network effect is particularly
evident in Europe, where South Korean weapons systems are being
integrated into NATO's collective defense structure.

North Korea's partnership with Russia represents a deepening of
an emerging authoritarian axis. This alignment provides both countries with strategic depth and capabilities they lack individually.
Russia gains access to North Korean ammunition and manpower,
while North Korea receives advanced technology and diplomatic
protection. The partnership has already demonstrated its impact on
international institutions, with Russia's veto of UN resolutions monitoring North Korean sanctions representing a fundamental break-
down in the post-Cold War consensus on non-proliferation
enforcement.

What began as Putin's attempt to emulate Catherine the Great and reshape Europe has inadvertently catalyzed transformations across Asia. Japan's break with post-war pacifist constraints, South Korea's emergence as a global defense supplier, and North Korea's projection of military power beyond its borders all represent fundamental shifts triggered by the Ukraine conflict.

These developments underscore that in an increasingly interconnected world, no conflict remains truly regional, and the consequences of aggression extend far beyond the immediate battlefield.

CHAPTER 4

INDIA AND THE UKRAINE WAR

It is possible to work closely with the Russian military including through joint production in India while Russia is waging an aggressive war against Ukraine only if one is prepared to separate ethics from statecraft and to hard-prioritize national security over value alignment.

That, in essence, is what New Delhi has done: it condemns the war in abstract terms, but refuses to dismantle the defense relationship that underpins its deterrent posture against China and Pakistan.

This approach represents one of the most significant strategic dilemmas facing contemporary India, testing the limits of its long-cherished principle of strategic autonomy against the moral imperatives of the international rules-based order.

And India's military relationship with Russia to support its defense capability against China is ironic in light of how China is subordinating Russia to its interests and one wonders when China turns the screws on with regard to this relationship. India and Russia form the two continental powers which China is working to reshape to its interests as Global China,

THE 2025 PUTIN VISIT AND DEFENSE "REORIENTATION"

The December 2025 Putin-Modi summit formalized a shift that had been underway for a decade: from Russia as India's dominant armory to Russia as an embedded defense-industrial partner inside India. The joint statement explicitly frames military and military-technical cooperation as "reorienting" toward joint R&D, co-development and co-production of advanced systems, calibrated to India's Atmanirbhar Bharat (self-reliance) agenda.[1]

The most concrete outcome is political endorsement of large-scale joint manufacturing of spare parts, components and other products for Russian-origin equipment in Indian service, under Make in India. These joint ventures are intended to meet Indian armed forces needs and enable exports to "mutually friendly" third countries, implying a shared interest in sustaining Russian-designed systems well beyond the Ukraine war. The summit also reaffirmed Russia's commitment to provide "uninterrupted shipments" of oil, gas, and coal to India, despite Western sanctions following Russia's invasion of Ukraine.[2]

This strategic partnership extends beyond immediate military needs. India and Russia finalized an economic cooperation program until 2030, aiming to boost annual bilateral trade to $100 billion.

The framework encompasses energy security, technology transfers in artificial intelligence and space exploration, labor mobility agreements, and expanded port and shipping cooperation. For Indian policymakers, these arrangements represent not merely transactional defense deals but a comprehensive recalibration of bilateral relations to serve India's long-term strategic interests in an increasingly multipolar world.

LEGACY DEPENDENCE AND THE STRUCTURE OF INDIA'S ARSENAL

India's choice is rooted in the structure of its force. A large share of India's combat aviation (Su-30MKI, MiG-29 variants), armor (T-72, T-90), air defense, submarines and even its lone aircraft carrier draw on Soviet/Russian designs, maintenance chains and training pipelines.[3]

The scale of this dependence is staggering: approximately 90% of the Indian Army's equipment, including the backbone of its tank fleet, is of Russian origin. Around 70% of the Indian Air Force's combat aircraft are Russian-designed, with the Su-30MKI fleet alone comprising 14 of its 30 squadrons. The Indian Navy's sole operational aircraft carrier, INS Vikramaditya, is a refurbished Soviet-era vessel, and its fighter complement consists entirely of Russian MiG-29K aircraft.[4]

Western and indigenous platforms are growing, but they are additive, not yet replacement-grade for the entire Soviet-era backbone of the order of battle. If New Delhi were to abruptly curtail military-technical cooperation with Moscow, the first-order effect would not be to stop Russian operations in Ukraine; it would be to compromise the readiness of Indian units facing a far more immediate adversary in the People's Liberation Army along the Line of Actual Control and a persistent Pakistani threat on the western front.[5]

Indian strategists thus treat Russian support as a "bridge dependency" they must manage down over a decade or more, not something they can switch off on moral grounds in 2025 without incurring acute operational risk.

This dependency has historical roots dating back to the Cold War era, when the Soviet Union emerged as India's primary arms supplier following U.S. embargoes and Western reluctance to transfer advanced military technology.

The relationship deepened through licensed production arrangements, joint ventures like BrahMos missiles, and technology transfers that have shaped India's defense industrial base for generations.

STRATEGIC AUTONOMY AND
MULTI-ALIGNMENT AS DOCTRINE

The intellectual framework for this choice is India's post-Cold War doctrine of strategic autonomy and, more recently, "multi-alignment."[6]

This represents the evolution of India's founding principle of non-alignment, adapted to contemporary geopolitical realities. Rather than see the system as binary, democracies versus autocracies, New Delhi seeks overlapping, issue-specific alignments: the Quad and deeper U.S. ties for Indo-Pacific maritime balancing; Russia (and to a lesser degree France) for legacy hardware, niche technologies and nuclear cooperation; Gulf partners for energy and remittances; and the "Global South" for normative leadership.

The only problem is that neither Russia nor China view it this way. How in fact are India's strategic interests challenged by the multi-polar authoritarian world in ways that simply Indian leaders refuse to see?

As External Affairs Minister Subrahmanyam Jaishankar articulated, this doctrine involves "engaging America, managing China, cultivating Europe, reassuring Russia, bringing Japan into play, drawing neighbors in, extending the neighborhood, and expanding traditional constituencies of support."[7]

This approach lets India argue that it is not "choosing" Russia over Ukraine, or Russia over the West, but choosing India over everyone else.

It keeps channels open with Moscow because, in Indian calculations, a Russia wholly dependent on China or more willing to arm Pakistan would be worse for Indian security than a Russia that still sees India as a privileged defense partner.[8]

The net result is a deliberate dual alignment: more interoperability and exercises with the U.S. and other democracies, coupled with sustained (but slowly diluted) dependence on Russian military-technical support. This multi-alignment strategy reflects India's assessment that rigid alliances in a multipolar world can be counter-productive, potentially drawing nations into conflicts that do not serve their interests while alienating potential partners.

The principle of strategic autonomy, rooted in India's colonial past and its founding commitment to sovereignty and independence, enables New Delhi to make foreign policy decisions free from external influence.[9]

This allows for the pursuit of national interests in an increasingly complex and multipolar world. During the Cold War, India was a founding member of the Non-Aligned Movement, maintaining neutrality in the U.S.-Soviet rivalry. This commitment to autonomy has continued to shape its global engagement, particularly evident during Prime Minister Narendra Modi's tenure, where India has demonstrated its ability to balance relationships across multiple power centers simultaneously.

THE QUAD

On paper, this sits awkwardly with India's participation in the Quad, a grouping rhetorically framed by Washington, Tokyo and Canberra as a partnership of democracies to uphold a rules-based order. New Delhi finesses that contradiction by insisting that the Quad is not an alliance, not aimed at any third country, and not a vehicle for democracy promotion or regime change but for "public goods" in the Indo-Pacific—maritime domain awareness, infrastructure, technology and resilient supply chains.

This semantic narrowing creates political room to maintain a values-lite narrative: India can sit with the U.S., Japan and Australia to talk about Chinese coercion in the South China Sea while sitting with Russia in formats like BRICS and the Shanghai Cooperation Organization, and while signing joint-production defense deals in New Delhi. For Indian policymakers, the Quad is a China-management instrument, not a civilizational bloc obliging them to sanction or isolate Moscow over Ukraine.[10]

This compartmentalization allows India to participate in Western-led mini-lateral forums while remaining deeply integrated within forums where Chinese and Russian interests play major roles. Despite periodic violent incidents like the 2020 Galwan Valley clash with China, India has managed to keep functional relationships with both competing powers.

This delicate balancing act reflects India's assessment that in a fragmenting global order, maintaining multiple strategic partnerships even with adversarial powers serves its national interests better than exclusive alignment with any single bloc.

INDIA'S STANCE ON THE UKRAINE WAR

Since February 2022, India has walked a narrow line on Ukraine. It has repeatedly invoked the UN Charter, territorial integrity and sovereignty; Modi's much-quoted remark that "this is not an era of war" to Putin in 2022 has been reiterated in various forums, including around the 2025 summit.

Yet at key UN votes condemning Russia's invasion or demanding withdrawal, India has mostly abstained, preferring language that calls for cessation of hostilities and dialogue without naming Moscow as the aggressor.

Economically, India has sharply increased imports of discounted Russian crude and fertilizer since 2022, arguing that as a large developing economy with millions in poverty it cannot sacrifice affordable energy and food security to Western sanctions regimes. India's oil imports from Russia surged from negligible levels before the war to making Russia one of its top suppliers, with bilateral trade reaching $68.7 billion in the last fiscal year.

Diplomatically, it has sent officials to both Western-backed Ukraine peace formats and Russian-convened fora, positioning itself as a potential mediator while avoiding any step that would be read in Moscow as alignment with the Euro-Atlantic position.

THE "SOVEREIGNTY PARADOX"

This is often described as India's "sovereignty paradox": a country that has long championed territorial integrity and decolonization effectively underwriting, through its neutrality and purchases, a major violation of those norms. Indian commentators respond that Western practice on sovereignty has been highly selective, from Iraq to Libya and Kosovo, and that a state so exposed to external pressure cannot afford a purist reading of norms that others themselves violate.

India does not deny that Russia has violated Ukrainian sovereignty. It simply refuses to make that admission the organizing principle of its foreign or defense policy.

Principles, in this view, are filters through which interests are pursued, not trump cards that override the imperative to deter China and manage a brittle internal development trajectory. This is a cold, realist argument, but it is widely shared in Indian strategic circles and across much of the non-Western world.

This position reflects a broader skepticism about Western moral authority that resonates throughout the Global South. Many developing nations view the selective application of international law by Western powers as undermining the credibility of calls for universal adherence to rules-based order.

For India, which experienced colonialism and has long advocated for a more equitable international system, this historical memory shapes its contemporary foreign policy calculations. The country's leadership argues that moral consistency requires acknowledging that major powers, Western and non-Western alike, have repeatedly subordinated principles to interests when vital national security concerns are at stake.

CO-PRODUCTION WITH RUSSIA

Working "closely with the Russian military" inside India through joint R&D, licensed manufacture and co-production is rationalized in Delhi on three grounds.

- First, localization of spares and systems reduces vulnerability to sanctions and supply disruption by giving India more control over the life-cycle of legacy Russian platforms. This became particularly urgent after Russia's invasion of Ukraine disrupted defense production and Western sanctions constrained Moscow's ability to fulfill export orders.
- Second, it is framed domestically as a way to accelerate technology absorption and industrial capacity under Atmanirbhar Bharat, using Russian designs as a scaffold while reducing long-term foreign dependence. Russia's willingness to transfer technology and establish joint ventures, exemplified by BrahMos missiles, licensed production of Su-30MKI fighters, and T-90 tank manufacturing, contrasts favorably with Western reluctance to share cutting-edge military technology. This technology transfer enables India to build domestic defense manufacturing capabilities that would otherwise take decades to develop independently.
- Third, officials argue that these arrangements do not materially change Russia's war-fighting capacity in Ukraine, since much of the production is for Indian consumption and for third states that are already outside Western sanctions coalitions. In other words, they claim a moral distinction between, say, supplying artillery shells to Russian forces in Donbas and manufacturing T-90 spares in India to keep Indian armored regiments operational on the LAC. To Western and Ukrainian observers that line is unpersuasive; to Indian planners, it is the difference between enabling aggression and maintaining an already-fielded deterrent.

THE COSTS AND LONG-TERM RISK

This balancing act is not cost-free.

India's stance on Ukraine has generated skepticism in European capitals and among parts of the U.S. policy community that otherwise see India as a natural democratic partner. Indian purchases of cheap Russian oil and continued arms cooperation have been portrayed in Western and Ukrainian media as financing the Kremlin's war machine, undercutting New Delhi's claim to be a principled voice of the "Global South."

There are also strategic risks. A Russia that emerges from the Ukraine war more beholden to China may have diminished incentive or capacity to prioritize Indian requirements; diversification and indigenization thus become urgent not just for moral reasons, but because Russia could be a less reliable partner going forward.

Russia's defense industry, severely disrupted by the war and Western sanctions, has struggled to maintain delivery schedules for existing contracts, including delayed shipments of S-400 air defense systems to India.

Over time, if Indian and Western defense ecosystems deepen enough through technology partnerships, joint production and shared ISR architectures, the relative value of the Russian tie will inevitably shrink, potentially making the current ethical compromise a time-bounded one.

IS THE POSTURE SUSTAINABLE?

The core bet in New Delhi is that time is on India's side. Indian policymakers aim to keep enough of the Russian relationship alive to avoid a near-term capability shock and a Russia-Pakistan or tighter Russia-China tilt, while steadily lowering the share of Russian origin in India's arsenal and moving more production onshore.

In parallel, they are expanding operational cooperation and technology ties with the U.S. and other democracies so that, in five to ten years, the security cost of a sharper distancing from Moscow will be much lower than it is today.

Whether this dual track. moral distance, material engagement, can be maintained without a triggering crisis is uncertain. A dramatic escalation in Ukraine or a Russian action directly impinging on Indian interests could force choices that Delhi is keen to avoid.

The International Criminal Court's warrant for Putin's arrest, while not enforced during his India visit, highlights the reputational costs India bears by hosting leaders accused of war crimes.

Additionally, Russia's increasing dependence on China for military components and economic support may gradually erode the independent pole that India values in Moscow, potentially undermining the strategic rationale for maintaining close ties.

For now, however, the answer to the central question is that India's leadership accepts the normative dissonance as the price of navigating a hostile neighborhood with an imperfect toolkit, and prioritizes the survival and autonomy of the Indian state over coherent alignment with the democratic camp's stance on Russia's war.

This approach reflects India's assessment that in an increasingly multipolar world characterized by great power competition, strategic autonomy even when it creates uncomfortable contradictions serves its national interests better than rigid ideological alignment.

The country's policymakers calculate that maintaining this delicate balance, while difficult and subject to criticism, provides the flexibility necessary to secure India's long-term security and development objectives in an uncertain global order.

India's handling of the Ukraine–Russia war illustrates how ad hoc balancing has so far substituted for a fully articulated rules-agenda. New Delhi has combined moral language ("this is not an era of war") with a refusal to join Western sanctions and continued purchases of discounted Russian oil, preserving a long-standing strategic and defence relationship it still sees as useful against China. This posture limits reputational costs in the West while signaling to Moscow that India will not help isolate it completely, thereby keeping Russia as a residual pole in Asia rather than allowing it to become wholly dependent on Beijing.

Yet this balancing act exposes a key conceptual gap. India resists both a Western-dominated and a China-dominated order, but has only begun to spell out what concrete, alternative rules it wants to see embedded in global regimes. Strategic autonomy in practice has meant flexibility on sanctions, voting patterns, and energy choices, but less sustained effort to codify principles that reconcile India's interest in sovereignty and territorial integrity with its reluctance to punish norm-breakers when core interests are at stake.

Without a more explicit normative framework, India risks looking merely transactional leveraging crises for short-term advantage rather than acting as a rule-shaper for a post-hegemonic order.

The unresolved project, therefore, is to move from reactive multi-alignment to proactive norm entrepreneurship that clearly constrains a Chinese-centric hierarchy without simply restoring Western primacy.

That would require India to champion specific rules, on territorial revisionism, economic coercion, debt sustainability, and digital governance, that can attract both Western and Global South support while remaining consistent with its own practices.

Unless New Delhi can define and socialize such rules, strategic autonomy will continue to be read as hedging between camps, rather than as an affirmative attempt to redesign the order in a way that protects India's security and development interests in the shadow of a rising China.

And I think notably the war in Ukraine is strengthening China significantly at the expense of Russia. How will India enjoy its own subordination to Global China and the new "rules of order."?

SECTION VI
THE U.S. DYNAMIC

CHAPTER 1
BIDEN'S UKRAINE STRATEGY

From the chaotic withdrawal from Afghanistan in August 2021 to the massive military aid packages delivered to Ukraine following Russia's full-scale invasion in February 2022, Biden's policy was characterized by tactical responsiveness coupled with persistent questions about strategic coherence.

The Biden administration's handling of Ukraine emerged in the shadow of Afghanistan, a withdrawal that severely damaged American credibility and influenced adversary calculations about U.S. resolve. As the administration scrambled to rally Western allies against Russian aggression, it focused on shaping NATO unity and providing substantial military support to Ukraine.

The Biden administration pursued an aggressive public diplomacy strategy in the lead-up to Russia's invasion. Biden engaged in robust diplomacy with NATO allies, Congress, and the American public, attempting simultaneously to deter Putin and prepare the world for worst-case scenarios. CIA Director William Burns traveled to Moscow in November 2021 to warn Putin directly about the consequences of invasion.

This public diplomacy effort represented a deliberate contrast with Afghanistan. The administration sought to project unity with

NATO partners, coordinate sanctions packages in advance, and establish clear consequences before any invasion occurred.

Following Russia's invasion on February 24, 2022, the Biden administration mobilized an extraordinary level of military assistance to Ukraine. In the first weeks after the invasion, Biden authorized an initial $800 million military aid package, including artillery weapons, howitzers, tactical drones, and ammunition.[1] This marked the beginning of what would become the largest foreign aid commitment to a European country since the Marshall Plan after World War II.

By the end of Biden's presidency, the United States had provided or committed over $66.5 billion in military assistance to Ukraine.[2] This included more than 1.6 million artillery rounds, tens of thousands of Javelin anti-tank missiles, HIMARS rocket systems, M1 Abrams tanks, Patriot air defense systems, and eventually, permission for European allies to supply F-16 fighter aircraft.

The assistance evolved over time as the administration gradually overcame escalation concerns. Initially hesitant to provide certain weapons systems for fear of provoking Russian retaliation, the administration incrementally expanded the types and capabilities of weapons provided. In 2024, it began supplying long-range ATACMS missiles that could strike targets nearly 200 miles away, and by November 2024, it lifted restrictions on using these weapons to strike inside Russian territory.[3]

As Biden's presidency drew to a close, the administration accelerated weapons deliveries, rushing military support to Ukraine before the anticipated policy changes under President-elect Donald Trump. After Trump's November 2024 election victory, the Biden administration announced multiple aid packages totaling hundreds of millions of dollars in final weeks, attempting to provide Ukraine with as much support as possible before the transition.[4]

STRATEGIC SUCCESSES

Perhaps the Biden administration's greatest strategic achievement was maintaining and strengthening NATO unity in the face of Russian aggression. The invasion of Ukraine produced what Putin likely did not anticipate: a revitalized and expanded NATO alliance. Finland and Sweden, historically neutral countries, applied for and received NATO membership, dramatically extending the alliance's border with Russia.

The administration successfully coordinated unprecedented economic sanctions against Russia, largely cutting it off from the global financial system. It also managed to maintain a unified Western response despite diverse national interests and varying levels of exposure to Russian energy supplies.

The military assistance provided by the United States and its allies prevented Ukraine from collapsing in the face of Russian invasion. When Russia launched its assault in February 2022, many observers expected Kyiv to fall within days. The combination of fierce Ukrainian resistance and Western military support defied these predictions, enabling Ukraine to repel the initial assault on Kyiv and launch successful counteroffensives in the fall of 2022 that recaptured significant territory, including the strategic city of Kherson.

One analysis concluded bluntly that "Ukraine would have lost the war without this additional U.S. funding."[5] The United States provided approximately half of all military aid to Ukraine, making it indispensable to Ukraine's ability to sustain its defense.

STRATEGIC SHORTCOMINGS

The Biden administration faced persistent criticism for lacking a clear strategic endgame. As early as March 2022, just weeks after the invasion, U.S. and European officials acknowledged they saw "no clear end to the military phase of this conflict." The administration's strategy, according to senior officials, was "to ensure that the economic costs for Russia are severe and sustainable, as well as to continue supporting Ukraine militarily in its effort to inflict as many defeats on Russia as possible."[6]

However, this approach begged fundamental questions:

- What would constitute victory?
- How would the conflict end?
- What were the conditions for a negotiated settlement?

President Biden himself articulated contradictory objectives at different times. In March 2022, he declared that Putin "cannot remain in power," suggesting regime change as an objective, only for the White House to walk back the comment. Defense Secretary Lloyd Austin stated that the goal was to see "Russia weakened to the degree that it can't do the kinds of things that it has done in invading Ukraine," while Biden promised support for "as long as it takes" without defining what "it" was or "how long" would be sufficient.[7]

Congress mandated that the administration produce an unclassified strategy document outlining specific objectives, national security interests, and metrics for measuring progress. The administration submitted a report, but it was entirely classified, frustrating both Congress and civil society organizations seeking clarity on American strategic objectives.[8]

The Biden administration found itself trapped in a strategic paradox or Catch-22 situation. On one hand, Biden pledged unwavering support for Ukraine "for as long as it takes." On the other hand, the administration remained deeply concerned about escalation and the prospect of direct confrontation with nuclear-armed Russia.[9]

These objectives increasingly worked at cross purposes. The administration's escalation concerns led it to delay or deny certain weapons systems, imposing restrictions on how and where weapons could be used. This incremental approach, while perhaps prudent from a risk-management perspective, prevented Ukraine from achieving decisive military advantages that might have created conditions for negotiations from a position of strength.

The Biden administration initially operated under the assumption that time would favor Ukraine and that Russian military and economic capacity would degrade under the twin pressures of sanctions and battlefield losses.

However, by 2024, this assumption appeared increasingly doubtful. U.S. and European defense officials acknowledged that Russia was able to reconstitute its military "far faster than initial estimates suggested," with its army described as actually "now larger by 15% than it was when it invaded Ukraine" according to testimony by Gen. Christopher Cavoli, commander of U.S. European Command, in April 2024.[10]

Despite some early contraction in 2022, Russia's economy posted notable growth in 2023 and 2024, confounding many predictions of collapse in the face of wide-ranging sanctions. Several analyses record Russian GDP growth of about 3.6% in 2023 and estimates ranging from 3.9% to 4.1% in 2024, with strong wage and investment trends reported. These economic results have allowed Russia to fund expanded military efforts and adapt to sanctions more effectively than expected.[11]

Meanwhile, Ukraine's strategic position became progressively more precarious. Starved of weapons and ammunition during congressional delays on aid packages, Ukraine was forced to cede ground, with Russia making its most significant advances since July 2022. The war became a grinding battle of attrition that favored the larger power with greater population reserves and industrial capacity.

By late 2024, polling showed that over half of Ukrainians believed their country should seek a negotiated end to the conflict, with a majority of those willing to make territorial concessions.[12] This represented a dramatic shift from the early war period and reflected growing exhaustion with the human and economic costs of the conflict.

One of the most contested aspects of Biden's Ukraine policy concerns potential diplomatic opportunities that may have been missed or insufficiently pursued. In March and April 2022, just weeks after the invasion, Turkey mediated negotiations between Russian and Ukrainian delegations in Istanbul.

Negotiator Oleksandr Chalyi stated: "We managed to find a very real compromise. We were very close in the middle of April, in the end of April, to finalize our war with some peaceful settlement"[13]

The draft agreement reportedly included Ukrainian neutrality in exchange for security guarantees. The talks collapsed for multiple reasons, including evidence of Russian war crimes at Bucha and opposition from Ukraine's Western partners.[14]

The deal faced a fundamental problem: it required Western powers to provide security guarantees that would obligate them to defend Ukraine militarily, but Ukraine had not consulted with Western allies during the negotiations, and Western powers were unwilling to make such commitments

The administration's approach to diplomacy remained cautious throughout the conflict. While maintaining channels of communication with Russia, it largely avoided pursuing active negotiations, citing the Kremlin's lack of seriousness about talks.

The administration's support for Ukraine's maximalist war aims, complete liberation of all occupied territory, including Crimea, created a tension between stated objectives and realistic capabilities. While rhetorically supporting these goals, U.S. military officials privately expressed skepticism about their achievability.

This gap between aspirations and assessments reflected a broader problem: the administration appeared reluctant to publicly acknowledge that Ukrainian and American interests were not identical. Ukraine, fighting for its survival and sovereignty, understandably sought the complete expulsion of Russian forces. The United States, however, had different priorities: safeguarding Ukraine's existence as a sovereign state, avoiding direct conflict with Russia, and managing the broader implications for European security and American commitments worldwide.[15]

However, the administration never clearly articulated this distinction, leading to continued uncertainty about American objectives and the conditions under which the United States would consider the conflict successfully concluded.

CONCLUSION

Biden's approach to Ukraine successfully prevented Ukrainian collapse, maintained unprecedented NATO unity, and imposed severe costs on Russia for its aggression. These achievements are substantial and may prove historically consequential.

However, the policy suffered from persistent strategic ambiguity about objectives, pathways to resolution, and acceptable outcomes. As the conflict entered its third year and a new U.S. administration took office, the fundamental questions about Biden's Ukraine policy remained unresolved.

- Had the substantial military aid been sufficient to achieve American strategic objectives?
- Was the policy sustainable over the long term?
- Could the approach have produced better outcomes with different tactical choices or greater strategic clarity?

History's judgment on Biden's Ukraine policy will ultimately depend on how the conflict concludes and whether the post-war settlement serves American interests in European security and the maintenance of international norms against aggression.

CHAPTER 2
TRUMP'S UKRAINE STRATEGY

I f Donald Trump's foreign policy approach can be characterized as "populist realism", combining aggressive domestic political messaging with pragmatic restraint in military commitments, then the war in Ukraine represents perhaps the most significant test of this framework.

The conflict embodies precisely the kind of complex, protracted foreign engagement that Trump has consistently criticized, while simultaneously presenting genuine strategic challenges that cannot simply be dismissed or ignored.

His approach to Ukraine reveals the tensions inherent in his foreign policy philosophy.

- How does a president who promised to end "endless wars" handle a conflict where American credibility and European security hang in the balance?
- How does someone who views international relationships through a transactional lens navigate a situation where democratic values and strategic interests intersect?
- And how does a leader who prioritizes domestic political

messaging manage a crisis that demands careful diplomatic coordination with allies?

The answer lies in understanding how Trump's characteristic patterns, the rhetoric-reality gap, strategic restraint, transactional diplomacy, and domestic-first communication, apply to this specific challenge.[1] His Ukraine policy reflects not a departure from his broader foreign policy approach, but rather its most consequential application.

THE 24-HOUR PROMISE: RHETORIC VERSUS REALITY

During his campaign, Trump repeatedly promised he could end the Ukraine war "in 24 hours," a claim that exemplifies his tendency toward dramatic statements designed for domestic political consumption rather than international diplomatic effect.

This promise served multiple political purposes: it signaled decisiveness to voters tired of protracted conflicts, implied that previous administrations had mishandled the situation, and suggested that Trump possessed unique negotiating abilities that could achieve rapid results.

Yet anyone familiar with the complexity of the Ukraine conflict, the territorial disputes, security guarantees, NATO membership questions, and fundamental disagreement over Ukrainian sovereignty, understands that no resolution could be achieved in 24 hours.

The promise was never meant as a literal diplomatic roadmap but rather as political messaging that resonated with Americans skeptical of continued involvement in foreign conflicts.

This rhetoric-reality gap mirrors Trump's approach to North Korea, where threats of "fire and fury" gave way to diplomatic summits, or his criticism of NATO allies that ultimately strengthened rather than weakened the alliance.

The pattern suggested that Trump's actual Ukraine policy would be more nuanced and pragmatic than his campaign rhetoric implied, focused on negotiated settlement rather than either complete disengagement or unlimited commitment.

STRATEGIC RESTRAINT AND THE LIMITS OF AMERICAN COMMITMENT

Perhaps the most defining characteristic of Trump's foreign policy approach has been his reluctance to commit American forces to extended overseas engagements. This preference for restraint shapes his approach to Ukraine in fundamental ways. While the United States has not deployed combat troops to Ukraine, the conflict has required sustained American military aid, intelligence support, and diplomatic engagement, precisely the kind of long-term commitment that Trump views skeptically. His criticism of "endless wars" in Afghanistan, Iraq, and Syria naturally extends to concerns about open-ended support for Ukraine, regardless of the conflict's different character and strategic implications.

This creates a tension between Trump's instinct for disengagement and the reality that abrupt withdrawal of American support could have catastrophic consequences for European security. The challenge for Trump is finding a middle path that limits American military and financial commitment while avoiding outcomes that would embolden adversaries and undermine allied confidence in American reliability.

His approach involves pushing for a negotiated settlement that ends active combat, even if it requires Ukrainian territorial concessions that many find unpalatable. This reflects his transactional view of foreign policy: rather than viewing the conflict through the lens of democratic values or long-term strategic positioning, Trump is more likely to focus on ending the immediate costs and risks of the conflict, even if the resulting settlement is imperfect.

This approach aligns with Trump's broader foreign policy pattern of prioritizing concrete, measurable outcomes. In this case, an end to active combat over broader ideological commitments to democratic values and international norms.

THE TRANSACTIONAL FRAMEWORK

Central to understanding Trump's Ukraine approach is recognizing his fundamentally transactional view of international relations. Just as he criticized NATO allies for insufficient defense spending and pressured them to increase their contributions, Trump views Ukraine support through the lens of burden-sharing and fair distribution of costs among allies.

From Trump's perspective, the United States has shouldered a disproportionate share of the cost of supporting Ukraine despite the conflict's primary impact on European security. While European nations have collectively provided substantial aid, Trump's transactional calculus focuses on the direct American financial commitment and the political costs of sustained involvement. His criticism resonates with American voters who question why the United States should bear such significant costs for a conflict that more directly threatens European than American interests.

This framing allows Trump to position himself as defending American interests against allies who have historically underfunded their own defense. His pressure on European nations to increase their support for Ukraine mirrors his broader approach to NATO burden-sharing, where public criticism and threats ultimately contributed to increased European defense spending.

The transactional approach extends beyond financial considerations to diplomatic strategy. Trump is likely to condition continued American support on European commitments to shoulder more of the burden, both in terms of military aid and in any eventual peace-keeping or security guarantee arrangements.

This creates leverage for negotiations with both European allies and with Russia, as Trump can argue that American support is not unlimited and that other parties must take greater responsibility for any settlement's implementation and enforcement.

THE CREDIBILITY PARADOX:
UNPREDICTABILITY AS STRATEGY

One aspect of Trump's foreign policy approach is what might be called the "credibility paradox" or the possibility that his unpredictable communication style and willingness to depart from conventional diplomatic practice could actually enhance American deterrence by creating uncertainty among adversaries about how the United States might respond to various provocations.

This dynamic plays out in complex ways regarding Ukraine.

- On one hand, Trump's criticism of unlimited support for Ukraine and his expressed admiration for Vladimir Putin in the past raise questions about American reliability and commitment to European security. Allies worry about whether they can count on American support, potentially undermining the cohesion necessary for effective resistance to Russian aggression.
- On the other hand, Trump's unpredictability could create deterrence effects that more conventional approaches lack. Putin cannot be certain how Trump might respond to particular Russian actions, whether Trump might authorize limited military strikes similar to Operation Midnight Hammer, dramatically increase military aid, or conversely push for immediate negotiations regardless of battlefield conditions. This uncertainty could make Putin more cautious about escalation or about rejecting reasonable settlement terms.

The credibility paradox extends to Trump's relationship with Ukrainian President Volodymyr Zelensky. Trump's willingness to pressure Ukraine to make concessions, something traditional diplomats might avoid for fear of appearing to abandon an ally, could actually create space for negotiations by signaling to both sides that continued conflict carries costs and that settlement terms will not perfectly satisfy either party.

Yet this same unpredictability carries risks. If allies cannot rely on American commitments, they may pursue independent arrangements that undermine collective security. If Putin believes Trump is desperate to exit the conflict regardless of terms, he may hold out for more favorable conditions. The credibility paradox cuts both ways, potentially enhancing deterrence while simultaneously raising questions about reliability.

DOMESTIC POLITICAL SUSTAINABILITY: THE HOME FRONT MATTERS

A key insight from Trump's foreign policy approach is his recognition that foreign policy requires domestic political sustainability. Unlike foreign policy establishments that sometimes pursue strategies with limited public support, Trump understands that foreign commitments cannot be maintained without ongoing public backing, particularly in a democracy where electoral shifts can produce dramatic policy reversals.

This insight applies directly to Ukraine. Public opinion polling has shown declining American support for unlimited aid to Ukraine, particularly among Republican voters who form Trump's political base. Many Americans question why the United States should continue bearing substantial costs for a conflict that seems to have no clear endpoint, particularly when domestic needs go unmet. Trump's skepticism toward open-ended Ukraine support reflects and reinforces these public attitudes.

This creates both constraints and opportunities. The constraint is that Trump faces political pressure from his base to limit American involvement, making sustained commitment difficult even if strategic considerations suggest otherwise. The opportunity is that a negotiated settlement, even an imperfect one, could be presented as fulfilling Trump's promise to end the conflict and avoid endless wars, generating political benefits regardless of the settlement's specific terms.

Trump's domestic-first communication style means his Ukraine messaging will be crafted primarily for American audiences rather than for diplomatic effect. He will frame policy decisions in terms of American interests, cost savings, and avoiding quagmires rather than in terms of democratic values or international norms. This messaging may alarm traditional foreign policy establishments but resonates with voters skeptical of foreign interventions.

The challenge is ensuring that domestic political considerations do not override strategic necessities. A settlement that plays well politically in the short term could have long-term consequences for American credibility and European security.

Yet Trump's approach suggests he believes previous administrations erred by prioritizing diplomatic establishment preferences over public support, leading to foreign policy approaches that lacked domestic sustainability and ultimately failed.

THE EUROPEAN DIMENSION: ALLIANCE POLITICS AND BURDEN-SHARING

Trump's approach to Ukraine cannot be separated from his broader relationship with European allies, characterized by public criticism paired with pressure for increased burden-sharing. His view is that European nations have historically underfunded their defense while relying on American security guarantees, and that the Ukraine conflict provides an opportunity to reset this relationship toward more equitable burden-sharing.

This creates both friction and alignment with European interests. The friction comes from Trump's transactional approach and public criticism, which can strain diplomatic relationships and create uncertainty about American reliability. European leaders accustomed to American foreign policy grounded in shared democratic values find Trump's approach unsettling, particularly when he appears willing to negotiate over Ukrainian sovereignty or to condition support on financial considerations.

Yet there is also alignment. Many European nations recognize they have underinvested in defense and that greater European strategic autonomy is necessary regardless of American political changes. Trump's pressure, while diplomatically uncomfortable, accelerates trends toward increased European defense spending and capability development that even traditional American foreign policy advocates view as necessary.

The key question is whether Trump's approach will produce greater European burden-sharing or instead lead to European nations pursuing independent arrangements that undermine collective security.

- If European allies respond to American pressure by taking greater responsibility for their own security, including Ukraine support, Trump's approach could strengthen rather than weaken European security architecture.
- If instead they lose confidence in American reliability and fragment into competing national approaches, the result could be strategic incoherence that serves neither American nor European interests.

THE RUSSIA CALCULATION: NEGOTIATING FROM STRENGTH

Despite Trump's past expressions of admiration for Vladimir Putin, his actual approach to Russia has combined elements of pressure and engagement. His administration provided lethal military aid to Ukraine, maintained sanctions, and criticized Russian actions while simultaneously pursuing diplomatic engagement. This pattern suggests his approach to negotiating over Ukraine will involve both pressure tactics and willingness to compromise that Trump views as pragmatic.

From Trump's perspective, ending the Ukraine conflict requires understanding Putin's core interests and identifying areas where accommodation is possible without compromising essential American interests. This transactional approach treats the conflict not as a morality play about democracy versus authoritarianism, but as a dispute over influence, territory, and security arrangements where negotiated settlement is possible if both sides receive sufficient benefits.

This creates risk of legitimizing Russian aggression and accepting territorial changes achieved through force, potentially encouraging future aggression. Yet Trump would argue that continued conflict serves neither American nor Ukrainian interests, and that a settlement acknowledging some Russian gains while preventing further conflict serves pragmatic American objectives better than open-ended commitment to reversing all Russian territorial gains.

The key is whether Trump can negotiate from sufficient strength to prevent Russia from simply consolidating gains and preparing for future aggression. This requires maintaining credible deterrence while demonstrating willingness to accept imperfect outcomes that end active combat. It also requires European commitment to any settlement's enforcement, ensuring that reduced American involvement does not create security vacuums that Russia can exploit.

Trump's unpredictability could serve him well in negotiations with Putin. If Putin cannot predict whether Trump might authorize military strikes similar to Operation Midnight Hammer in response to particular Russian actions, or might dramatically increase military aid if negotiations fail, he may be more willing to accept settlement terms that preserve Ukrainian sovereignty even if acknowledging some Russian territorial gains.

CONCLUSION: PRAGMATISM, NOT IDEALISM

Trump's approach to Ukraine reflects his broader foreign policy framework of populist realism: combining domestic political messaging with pragmatic pursuit of limited objectives, prioritizing tangible outcomes over ideological commitments, and seeking nego-tiated settlements rather than open-ended military engagements. While this approach creates legitimate concerns about American reliability and allied confidence, it also reflects recognition that foreign policy requires domestic political sustainability and that previous approaches failed to prevent or resolve the conflict.

The ultimate test of Trump's Ukraine policy will be whether he can achieve a negotiated settlement that ends active combat without creating conditions for renewed aggression, maintains allied confi-dence in American reliability, and secures sufficient burden-sharing from European allies to make any settlement sustainable. This requires threading a difficult needle between disengagement and unlimited commitment, between accommodating Russian interests and defending Ukrainian sovereignty, and between satisfying domestic political pressures and maintaining strategic credibility.

The question is not whether his approach perfectly serves demo-cratic values, but whether it serves American interests better than alternatives. That debate will define not just Ukraine policy, but the broader question of America's role in the world during Trump's second term.

CHAPTER 3
COMPARING THE TWO APPROACHES

T he war in Ukraine has become the defining foreign policy challenge of the 2020s, exposing fundamental disagreements about America's role in the world. Biden's approach represents the continuation of post-Cold War values-based internationalism, while Trump's framework signals a fundamental departure from this approach.

Understanding these competing visions is essential not only for predicting the future trajectory of American support for Ukraine, but for grasping the broader transformation underway in American foreign policy.

BIDEN'S APPROACH: VALUES-BASED INTERNATIONALISM UNDER PRESSURE

Biden's Ukraine policy cannot be understood without first examining the catastrophic withdrawal from Afghanistan in August 2021. The withdrawal raised fundamental questions about American commitment and competence.

Following Russia's invasion on February 24, 2022, the Biden administration mobilized an extraordinary level of military assistance to Ukraine. By the end of Biden's presidency, the United States had provided or committed over $66.5 billion in military assistance, the largest foreign aid commitment to a European country since the Marshall Plan after World War II.

The assistance evolved over time as the administration gradually overcame escalation concerns. Initially hesitant to provide certain weapons systems for fear of provoking Russian retaliation, the administration incrementally expanded the types and capabilities of weapons provided.

The military assistance provided by the United States and its allies prevented Ukraine from collapsing. When Russia launched its assault in February 2022, many observers expected Kyiv to fall within days. The combination of fierce Ukrainian resistance and Western military support defied these predictions, enabling Ukraine to repel the initial assault on Kyiv and launch successful counteroffensives in the fall of 2022 that recaptured significant territory.

Despite tactical successes, the Biden administration faced persistent criticism for lacking a clear strategic endgame. The Biden administration found itself trapped in a strategic paradox. On one hand, Biden pledged unwavering support for Ukraine 'for as long as it takes.' On the other hand, the administration remained deeply concerned about escalation and the prospect of direct confrontation with nuclear-armed Russia. These objectives increasingly worked at cross purposes.

TRUMP'S APPROACH: POPULIST REALISM AND STRATEGIC RESTRAINT

While his predecessors carefully calibrated their language for international audiences, Trump's approach to diplomacy has been shaped primarily by domestic political considerations, creating what appears to be a fundamental contradiction: hard-hitting rhetoric paired with surprisingly restrained actions, particularly regarding military intervention.

And Trump II is not a complete continuation of Trump I. Trump 2 is best understood as an effort to turn the first term's insurgent "America First" impulses into a more structured project of statecraft, focused on advantage in a multipolar world and control of the near region. It retains the populist critique of globalization, migration, and progressive culture but now couples it to a more deliberate use of tariffs as industrial policy, tighter presidential control over the bureaucracy, and more systematic use of federal levers and red-state partnerships to build durable conservative institutions. In this sense, Trump 2 is less about withdrawal than about restructuring U.S. commitments around perceived national benefit and rebuilding domestic capacity for long-run strategic competition.

In foreign and defense policy, Trump 2 wraps continued NATO deterrence, Ukraine assistance, and forward presence in Europe in a language of sovereignty, border security, and hemispheric focus that alarms traditional Atlanticists but does not actually abandon European security. The 2025 National Security Strategy and National Defense Authorization (NDAA) together preserve Article 5, lock in multi-year support for Ukraine, and constrain force cuts in Europe, even as they signal that Europeans must shoulder more of the conventional burden.

CONTRASTING PHILOSOPHIES: LEGACY POLICY VS. NEW PARADIGM

Biden's approach clearly continues post-Cold War American foreign policy traditions emphasizing democratic values, alliance cohesion, and rules-based international order. Yet the policy ultimately suffered from strategic ambiguity, a commitment to Ukrainian victory while fearing escalation, leading to incremental support that may have prolonged conflict without enabling decisive outcomes.

Trump's approach fundamentally breaks with decades of American foreign policy consensus on several fronts.

- First, his foreign policy rhetoric is crafted primarily for American domestic political consumption rather than

international audiences, treating diplomatic communication as another venue for populist messaging.
- Second, his transactional framework views international relations through cost-benefit analysis rather than shared values, demanding explicit reciprocity from allies.
- Third, Trump demonstrates genuine strategic restraint and reluctance toward extended military commitments, contrasting sharply with post-9/11 interventionism.
- Fourth, he publicly pressures allies while demanding increased burden-sharing, challenging the traditional approach of private diplomatic persuasion.
- Finally, Trump shows willingness to accept imperfect negotiated settlements that end immediate conflicts rather than pursuing maximalist objectives based on principle.

This represents not merely tactical differences but a fundamental reorientation of American foreign policy priorities. Where traditional approaches emphasized maintaining alliances and upholding international norms even at significant cost, Trump's framework prioritizes limiting American commitments and achieving concrete, measurable outcomes that can be justified to domestic audiences.

TWO COMPETING VISIONS

The stark contrast between Biden's and Trump's approaches to Ukraine reflects fundamental disagreements about America's role in the world that extend far beyond this single conflict.

Biden represents evolution within traditional American internationalism. Yet Biden's policy struggled to translate tactical success into strategic resolution, caught between commitment to Ukrainian victory and fear of escalation. The result was incremental support that may have prolonged conflict without enabling decisive outcomes, alongside strategic ambiguity about ultimate objectives that frustrated both supporters and critics.

Trump represents a fundamental departure from this tradition, prioritizing domestic political sustainability over alliance solidarity, pursuing transactional burden-sharing over values-based partnerships, and seeking pragmatic settlements over ideological commitments. His approach promises to limit American exposure and achieve concrete outcomes that can be justified to war-weary voters, reflecting recognition that foreign policy requires domestic political support to be sustainable.

Whether Trump's approach produces sustainable outcomes or undermines American credibility remains the critical question.

- Can he negotiate a settlement that ends active combat without creating conditions for renewed aggression?
- Can he pressure European allies to assume greater responsibility without fragmenting collective security?
- Can he balance domestic political imperatives with strategic necessities?

The ultimate test will be whether either approach can achieve what the other could not: a resolution to the conflict that serves American interests while avoiding the pitfalls that have plagued previous policies. The answer will shape not only Ukraine's future, but the broader trajectory of American power and influence in the 21st century.

SECTION VII
LOOKING TOWARDS 2026

CHAPTER 1
THE STATE OF PLAY: OCTOBER 2025

The maneuvers in the war in Ukraine in the Fall of 2025 show several dynamics at once. The Russians are using attacks on civilian targets to undercut the will of the Ukrainian people to fight. But Putin has a significantly constricting economy and the threat of the Russian people becoming tired of the war as they did in Afghanistan. The Ukrainians are ramping up their attacks on Russian military and civilian infrastructure to erode the support for the war in Russia by bringing it home to the Russian people.

It is clearly a war entering a key phase. As the war in Ukraine enters its fourth autumn, the conflict has reached a critical inflection point. The fighting that began with Russia's full-scale invasion in February 2022 has evolved into a complex, multi-dimensional struggle that now directly threatens to draw NATO into its first military confrontation with Russia since the alliance's founding. The dynamics unfolding in Fall 2025 reveal a war that is simultaneously grinding forward on the battlefield while escalating dangerously in the diplomatic and strategic spheres.

Despite more than three and a half years of fighting, Russia continues to make territorial gains in eastern Ukraine, though at a pace that belies the enormous costs Moscow has incurred. According to data compiled by the Institute for the Study of War, Russian forces gained approximately 226 square miles of Ukrainian territory in the four-week period from August 19 to September 16, 2025. While this represents a slight decrease from the previous four-week period, the trend since the beginning of 2025 shows an average monthly Russian gain of 169 square miles.[1]

These territorial advances, concentrated primarily in the Donetsk region, tell only part of the story. Russian forces have captured numerous villages and towns, including areas near strategically important cities like Pokrovsk, Kostiantynivka, and Kupiansk. The advance has been characterized by intense artillery bombardment, massive use of drones, and grinding infantry assaults that consume enormous amounts of military equipment and personnel.

The human cost of these gains has been staggering. According to an April 2025 estimate by then-Supreme Allied Commander Europe General Cavoli, Russia has suffered more than 790,000 killed or injured, with an additional 50,000 missing.[2] Ukrainian casualties, while lower, are also severe, with President Volodymyr Zelenskyy estimating 400,000 killed or injured and 35,000 missing as of January 2025.[3] These figures represent a catastrophic loss of life that continues to mount with each passing month.

The material losses are equally devastating. Russia has lost over 22,000 tanks and armored vehicles according to open-source intelligence tracking, while Ukraine has lost nearly 10,000. The Defense Intelligence Agency estimated in May 2025 that Russia has lost at least 10,000 ground combat vehicles since the war's start, including more than 3,000 tanks, as well as nearly 250 aircraft and helicopters and more than 10 naval vessels.[4]

Both Russia and Ukraine have intensified attacks on civilian and military infrastructure throughout 2025, each attempting to erode the other side's capacity and will to continue fighting. This mutual targeting of infrastructure represents a deliberate strategy by both sides to bring the consequences of war home to their opponent's population.

Russia has maintained its intense campaign against Ukraine's energy infrastructure since late 2022, inflicting severe damage on the sector. By the end of 2024, Ukraine's available electricity generating capacity had reportedly shrunk from a prewar total of 56 gigawatts to about 9 gigawatts, with 64 percent of its 25 gigawatts of generation capacity either destroyed or located in territories under Russian occupation. As of September 2024, Russia's strikes had eliminated 80 percent of Ukraine's thermal capacity, making the country dependent on the three remaining Soviet-era nuclear power plants for roughly two-thirds of its power supply.[5]

The danger extends to nuclear facilities themselves. In October 2025, the International Atomic Energy Agency reported that the Russian-controlled Zaporizhzhia Nuclear Power Plant had been without offsite power for six days after recent attacks near the site. IAEA Director General Rafael Grossi expressed grave concerns about the situation, as the lack of power is needed to cool nuclear reactors and prevent a potential meltdown. Ukrainian President Zelenskyy warned that Russia's actions around the Zaporizhzhia plant represent "a threat to everyone," noting that "no terrorist in the world has ever dared to do with a nuclear power plant what Russia is doing now."[6]

Ukrainian drone strikes have targeted Russian oil refining capacity and military infrastructure deep inside Russian territory, particularly intensifying in August and September 2025. Specifically, Reuters and the *Kommersant* business daily, as cited by *The Moscow Times*, confirm that drone strikes reduced Russia's oil refining capacity by about 16–17 percent, equivalent to shutting down approximately 1.1 million barrels per day, during this period. The BBC and the analytics group Ciala provide additional context, noting that up to 38% of refining capacity was offline at peak interruption, but the figure directly attributable to drone strikes for late August and September is about one-quarter (roughly 16–17%) of the country's total capacity. This has caused gasoline shortages, rationing, and even temporary closures at hundreds of filling stations across Russia.[7]

These attacks have had tangible effects on Russian-occupied territories. In Crimea, which Russia annexed from Ukraine in 2014, fuel shortages resulting from Ukrainian drone strikes on oil refineries forced authorities to freeze fuel prices and impose petrol rationing, limiting motorists to purchasing 30 liters of fuel at a time.[8]

The attacks on civilian areas continue as well. A massive Russian attack on September 28, 2025, lasting more than 12 hours, unleashed close to 600 drones and dozens of missiles across seven Ukrainian regions, killing at least four people and injuring more than 70 others. In the Sumy region, a Russian drone attack killed a family of four, including two young children.[9]

While Russia continues its military operations, it faces mounting economic difficulties that threaten the sustainability of its war effort. Russia's economy is experiencing what one analysis described as "a slow train wreck," suffering from high inflation, prohibitive interest rates, a growing budget deficit, labor shortages, and unsustainable military spending.[10]

President Vladimir Putin signed a decree in October 2025 ordering the conscription of 135,000 men for military service, with Russian men aged 18 to 30 to be drafted between October 1 and December 31. This ongoing need for conscription highlights both the enormous personnel losses Russia has sustained and the difficulty of maintaining force levels necessary to sustain offensive operations.

The economic strain extends beyond military expenditures. Russia has had to develop increasingly complex sanctions evasion networks, partnering with countries like India, the United Arab Emirates, Turkey, and using shell companies in unexpected locations.

Despite these economic pressures, it remains unclear whether domestic discontent in Russia has reached levels that would threaten Putin's grip on power. The Kremlin has maintained strict control over information and dissent, making it difficult to assess the true state of public opinion regarding the war.

The most dangerous development in Fall 2025 has been a dramatic escalation in Russian violations of NATO airspace, creating the potential for direct military confrontation between the alliance and Russia. These incidents have reached an unprecedented scale and put NATO members on edge about Russian intentions.

The most significant incident occurred on September 10, 2025, when approximately 20 Russian drones violated Polish airspace during a Russian attack on Ukraine. Some drones were shot down by NATO jets, while others crashed on their own. This marked the first direct military engagement between NATO and Russia since the start of the full-scale invasion of Ukraine. The operation involved Polish F-16s, Dutch F-35s, Italian AWACS aircraft, NATO Multi Role Tanker Transport, and German Patriot systems.[11]

Less than two weeks later, on September 19, three armed Russian MiG-31 fighter jets violated Estonian airspace, remaining there for over 10 minutes. This prompted Estonia to invoke Article 4 of the NATO Treaty, which allows members to request consultations when they believe their territorial integrity or security is threatened. NATO's response was swift, with Allied aircraft scrambled to intercept and escort the Russian jets from Estonian airspace.[12]

Additional violations have been reported across NATO's eastern flank. Norway announced that Russia had violated its airspace three times in 2025 alone, twice over the sea near Vardø in Norway's far northeast, and once over an uninhabited area in the northeastern county of Finnmark. Romania and Latvia also reported single Russian drone violations of their airspace during September.[13]

Following the Estonian incident, NATO issued a strong statement condemning Russia's "dangerous violation" and warning that "Russia should be in no doubt: NATO and Allies will employ, in accordance with international law, all necessary military and non-military tools to defend ourselves and deter all threats from all directions."

NATO Supreme Allied Commander Europe General Alexus Grynkewich acknowledged the complexity of the situation, noting that shooting down unmanned drones carries different risks than engaging manned aircraft. "Shooting down manned aircraft like fighter jets clearly carries a higher risk of escalation if there's an engagement that kills someone on either side," Grynkewich explained. He acknowledged that different NATO nations have varying expectations about when and how NATO assets should inter-

vene, with some countries like Poland wanting "a very broad application" while "other nations make different judgments."[14]

The incidents have raised fundamental questions about Russian
motives. Russia is clearly probing NATO's defenses to identify weak
points and exploit fissures in the alliance's response. Estonian
Defense Minister Hanno Pevkur suggested that Russia may be trying
to force NATO to divert air defense resources from Ukraine to
defending its own territory: "Maybe their calculation was that now
the European countries have to send something additionally to
Estonia regarding the air defense assets, and that means they cannot
send it to Ukraine."[15]

The NATO airspace violations have revived discussions about
"closing the skies" over Ukraine, a proposal that was rejected early in
the war due to fears of direct NATO-Russia confrontation. While
NATO air patrols over Ukrainian territory have not been implemented, the concept is receiving renewed attention as Russian
airspace provocations continue.

Some proposals involve creating what has been termed an "Integrated Air and Missile Protection Zone" over western and central
Ukraine. Defense analyst Margaryta Vdovychenko has argued that
such a zone would not only restore stability and enable economic
recovery, but also allow Ukraine's Air Force to focus on defending the
eastern front. She estimates that implementing this would require
around 120 modern fighter jets, backed by early warning aircraft,
tankers, and robust intelligence, surveillance, and reconnaissance
capabilities.[16]

Poland is moving to amend its law on overseas military deployments to allow Polish forces to shoot down Russian drones over
Ukraine without prior approval from NATO or the EU. The Polish
Ministry of Defense introduced the draft legislation in June 2025,
aiming to restore rapid and independent response powers that had
been restricted by a law passed in 2022.

That earlier law required obtaining prior consent from NATO, the
EU, and the host country for Polish military operations abroad. This

change had been criticized for limiting Poland's ability to act against airborne threats like Russian drones near its border. The new, expedited bill seeks to remove these restrictions, permitting Poland's military to engage Russian unmanned aerial vehicles over Ukrainian airspace without international approval.[17]

European nations have increased their support for Ukraine throughout 2025, though responses remain uneven across the continent. Germany has pursued significant rearmament efforts. German Defense Minister Boris Pistorius described Russia as "the most significant and direct threat to NATO," pledging that Germany is ready to protect the Baltic region and will respond to Russian threats in a united and responsible manner.[18]

Lithuania's Defense Minister Dovile Sakaliene stated that Russia's recent airspace violations showed that NATO must move from "air policing missions" to "genuine air defense." The Danish government announced that production of solid rocket fuel for the Ukrainian FP-5 Flamingo missile would start in Denmark from December 1, 2025. British Prime Minister Keir Starmer visited Kyiv in early 2025 and announced £4.5 billion in aid, including 150 artillery barrels made in the UK and 15 additional Gravehawk air defense systems.[19]

European leaders are actively advancing plans to use frozen Russian assets held in European banks to support Ukraine, including at a recent summit in Copenhagen where both the proposal for a €140 billion loan (approximately $165 billion) backed by these assets and the concept of a "drone wall" were debated.

At the summit, European Commission President Ursula von der Leyen and leaders from Finland, Sweden, Denmark, and other nations voiced support for employing immobilized Russian central bank assets as collateral for the loan, which Ukraine would only repay if Russia fulfills its reparation obligations. The idea was first floated by von der Leyen and quickly gained traction as U.S. aid for Ukraine waned, putting increased financial responsibility on European states. Several leaders, including the Belgian Prime Minister, noted the need for legal clarity and risk-sharing among EU states, given many of these assets are held in Belgium.

The summit, dubbed the "drone wall" meeting, provided the first opportunity for EU leaders to formally debate the plan to use frozen Russian assets to fund the €140 billion ($165 billion) loan for military and economic support to Ukraine, alongside new defense initiatives like the drone wall and Eastern Flank Watch to counter Russia. The proposal received broad support, with leaders agreeing to further develop legal and financial frameworks before a final decision at a follow-up summit in three weeks.[20]

Nonetheless, Hungary and Slovakia refused to endorse European Union plans to phase out Russian gas deliveries by 2027-2028, particularly concerning the Turkish supply route, citing risks to their national energy security and economic interests. Both governments publicly criticized Brussels' push, arguing such measures would increase energy costs and undermine their ability to guarantee stable supplies.

Hungarian Prime Minister Viktor Orbán held a phone call with President Donald Trump in September 2025 specifically addressing Central European energy security. Orbán explained to Trump that Hungary's energy supply cannot be guaranteed without Russian gas and oil. Trump acknowledged this concern during or after the call, even as he pressed Hungary and other NATO allies to stop purchasing Russian energy to weaken Russia's war economy.[21]

The dynamics unfolding in Fall 2025 reveal a war that has entered what can best be described as an "escalatory stalemate with heightened NATO-Russia tensions." This phase is characterized by several simultaneous and potentially contradictory trends:

- Continued Russian territorial advances: Despite enormous costs, Russia maintains the operational initiative on the ground in eastern Ukraine, making slow but steady gains that gradually erode Ukraine's defensive positions.

- Intensifying infrastructure warfare: Both sides are escalating attacks on civilian and military infrastructure, attempting to break the other's will to continue fighting by bringing the war home to their opponent's population.
- Economic pressure mounting: Russia faces significant economic difficulties that threaten the long-term sustainability of its war effort, though whether these pressures will translate into meaningful constraints on Russian military operations remains unclear.
- Unprecedented NATO-Russia tensions: Russian airspace violations have created the most dangerous direct confrontation between NATO and Russia since the Cold War, with significant potential for miscalculation or escalation.
- European commitment deepening but divided: While European support for Ukraine has increased, significant divisions remain within Europe about how far to go in supporting Ukraine and how to respond to Russian provocations.
- Nuclear risks persisting: The situation at the Zaporizhzhia nuclear plant and Russia's general approach to targeting infrastructure create ongoing risks of nuclear accident, even absent intentional use of nuclear weapons.

This combination of factors creates a critical inflection point. a moment when the conflict could move in dramatically different directions depending on decisions made by key actors. The war is neither frozen into a stable stalemate nor rapidly moving toward resolution. Instead, it exists in a dangerous state where both breakthrough and catastrophic escalation remain possible, sometimes simultaneously.

The war in Ukraine as it stands in Fall 2025 represents perhaps the most dangerous moment since the initial invasion. The combination of ongoing military pressure, escalating infrastructure attacks, heightened NATO-Russia tensions, shifting U.S. policy, and mounting but uncertain economic pressures creates a highly volatile situation.

What is clear is that the war has reached a critical juncture. The decisions made in the coming months by leaders in Washington, European capitals, Kyiv, and Moscow will likely determine not just the outcome of this conflict, but the broader security architecture of Europe.

CHAPTER 2
RUSSIAN OPTIONS

I n October 2025, Russian President Vladimir Putin faces a confluence of crises that would have seemed unthinkable when his forces invaded Ukraine in February 2022.

Nearly four years into what the Kremlin euphemistically calls its "special military operation," Russia finds itself trapped in a war of attrition it cannot win militarily, cannot afford economically, and cannot abandon politically.

The strategic calculus that once seemed to favor Moscow based on leveraging superior resources and population against a smaller neighbor has been fundamentally upended by Ukrainian resilience, Western support, and Russia's own systemic vulnerabilities. In other words, by the nature of the global coalition of democracies backing Ukrainian.

The situation confronting Putin in autumn 2025 represents perhaps the most precarious moment of his presidency. On the battlefield, Russian forces have failed to achieve significant territorial gains despite catastrophic casualties. Ukraine's systematic campaign against Russian oil infrastructure has triggered the worst fuel crisis in post-Soviet history.

Meanwhile, Russian drone incursions into NATO airspace have

brought the alliance closer to direct confrontation with Moscow than at any point since the war began.

Each of these crises would be manageable in isolation, but their convergence creates a strategic trap from which there appears to be no clean escape.

- What seem to be Putin's limited options as of October 2025 in pursing his war in Ukraine?
- And with the war going from being a contained regional conflict to one pitting the authoritarian versus the democratic powers, how have his options changed?

THE MILITARY STALEMATE: PYRRHIC GAINS AT CATASTROPHIC COST

By October 2025, the Russian military offensive in Ukraine has devolved into a brutal war of attrition that bears uncomfortable parallels to the trench warfare of World War I. Despite Putin's March 2025 declaration that "there are reasons to believe we can finish off Ukrainian forces," and expectations of a major summer offensive, Russian forces have failed to achieve any significant breakthroughs.[1]

Key strategic objectives like the city of Pokrovsk remain firmly under Ukrainian control, while Russia's planned expansion into Sumy and Kharkiv oblasts has yielded only modest gains at an extraordinary human cost.

The price of these limited territorial advances has been staggering. Russia has been losing an estimated 100-150 troops per square kilometer of gained territory in 2025. Total Russian casualties, killed, wounded, missing, or captured, are nearing one million, a number that is rapidly approaching the combined total of all Russian and Soviet wars since World War II. This makes Ukraine Russia's second-deadliest conflict in a century, surpassed only by the Great Patriotic War against Nazi Germany.[2]

Yet these immense costs have not fundamentally altered the Kremlin's strategic calculus. As analysts at the Center for Strategic and International Studies note, what is driving Putin's approach to the war is not a focus on Russia's national interests or improving citizens' lives, but rather his view of the conflict as "the latest stage of the century-long Soviet and Russian struggle against the West," essential to cementing his historical legacy.[3] This ideological commitment means that rational economic considerations take a back seat to Putin's determination to avoid what he would perceive as a strategic defeat.

THE ENERGY CRISIS: UKRAINE STRIKES AT RUSSIA'S ECONOMIC JUGULAR

Perhaps the most dramatic development of 2025 has been Ukraine's systematic campaign to cripple Russia's oil refining capacity, a strategy that has proven devastatingly effective and is now causing cascading economic consequences across the Russian Federation.

Beginning in August 2025, Ukraine sharply increased its drone attacks against Russian oil refineries, striking at least ten sites and inflicting damage on refineries with a combined capacity representing around 17% of Russia's total processing ability, about 1.1 million barrels per day. These strikes led to acute fuel shortages, long lines, and price surges at gas stations in Crimea and across southern and far eastern Russia as outages became common. In September, the International Energy Agency reported that Russian revenues from oil product exports had dropped to levels last seen five years ago, linking this decline directly to the intensified campaign against refineries and highlighting its role in Russia's deepening economic slowdown.[4]

The situation deteriorated further in September when four major refineries were forced to halt operations after drone attacks. These included the Kirishi "Kinef" plant in Leningrad region, Russia's second-largest refinery, and Rosneft's Ryazan refinery, which ranks among the country's top five facilities.[5] By late September, nearly 38% of Russia's oil refining capacity, around 338,000 tons of crude per day, was offline.[6]

The domestic impact has been severe and politically sensitive. Gasoline output dropped by 1 million tons in September, leaving a supply shortfall equivalent to about 20% of domestic consumption.[7] More than 20 regions across Russia, from Sakhalin to Nizhny Novgorod, are now facing shortages, with the Far East and Crimea hit hardest. Gas stations in affected areas have limited sales to no more than 30 liters per customer, and long queues have become a source of public anger.

The Russian government has adopted a series of emergency measures in response to a worsening domestic fuel crisis triggered primarily by Ukrainian drone attacks on oil refineries which have taken nearly 40% of the country's refining capacity offline. Moscow has extended its ban on gasoline exports until the end of 2025, applying it to all exporters including producers, and imposed partial restrictions on diesel exports for non-producer resellers.

To alleviate the shortages, Russia has temporarily eliminated import duties on gasoline, diesel, and jet fuel within the Eurasian Economic Union to encourage imports. In a marked reversal for a nation long proud of its energy self-sufficiency, Russia is now preparing to import gasoline from China, South Korea, Singapore, and its traditional ally Belarus. These imports help offset growing domestic shortages and stabilize the fuel market amid ongoing refinery disruptions.[8]

DANGEROUS ESCALATION: TESTING NATO'S RESOLVE

As Ukraine's refinery strikes have intensified, Russian forces have engaged in increasingly dangerous provocations along NATO's eastern borders, raising the specter of direct confrontation between nuclear-armed powers.

The most serious incident occurred on the night of September 9-10, 2025, when 19 to 23 Russian drones violated Polish airspace during a large-scale attack on Ukraine. This unprecedented incursion triggered a Quick Reaction Alert, with Polish F-16 fighter jets, Dutch F-

35s, Italian AWACS surveillance aircraft, and German Patriot systems responding in a coordinated NATO operation. Up to four drones were confirmed shot down, marking the first time NATO forces had fired shots since the start of the Ukraine war.[9]

The Poland incident was not isolated. Three Russian MiG-31 combat aircraft entered Estonian airspace for 12 minutes in transit toward Kaliningrad on September 20, and a Russian drone breached Romanian airspace on September 13 which was the eleventh such incident according to Romanian Ministry of Defense data.[10]

NATO Secretary General Mark Rutte characterized these violations as "reckless" and "unacceptable," noting that Russian incursions into NATO airspace were "increasing in frequency."[11]

In response to this pattern of violations, NATO announced Operation Eastern Sentry on September 12, 2025, a comprehensive air defense mission for the alliance's eastern flank involving military forces from Denmark, France, the United Kingdom, Germany, and other nations. The European Union also backed plans for a "drone wall" along its borders with Russia and Ukraine, seeking to bolster defenses after the spate of airspace incursions exposed NATO's vulnerability to cheap, highly lethal drone technology.[12]

Russia's official response has been dismissive. The Kremlin claimed it had not targeted Polish territory and that the drones used in Ukraine have a flight range of no more than 700 kilometers, making Polish targets impossible to reach.[13]

The strategic intent behind these incursions appears to be testing NATO's unity and resolve. The calculated risk is that NATO's response will be strong enough to deter but not so aggressive as to trigger further Russian escalation.[14]

THE ECONOMIC TRAP:
MILITARIZATION WITHOUT EXIT

Underpinning all of Russia's strategic dilemmas is a fundamental economic problem: the country has become trapped in a war economy that cannot easily transition back to civilian production without triggering economic collapse yet cannot be sustained indefinitely without courting disaster.

By the first half of 2025, 67 of Russia's 89 regions reported severe budget deficits, with industrial hubs and resource-rich regions facing especially steep declines.[15] The Kremlin spent 8.5 trillion rubles (approximately $100 billion) on the military in just the first six months of 2025, even as sanctions and falling profits from natural resource exports eroded Russia's traditional income base. Oil and gas revenues fell by 16.9% in the first half of 2025 to 4.74 trillion rubles due to lower revenues from energy commodity sales.[16]

The economic growth figures, while superficially positive, mask a deeper malaise. Russia's GDP grew by 4.1% in 2024, but by the first quarter of 2025, annual growth had slowed to just 1.4% year-on-year, representing a 0.6% contraction compared to the previous quarter, the first quarterly contraction since the second quarter of 2022. Economic forecasts have been repeatedly downgraded, with the government now projecting growth of only 1.3% in 2026, down from earlier projections of 2.5% for 2025 and 2.4% for 2026.[17]

Sanctions, high taxes, and limited access to capital have created significant pressure on Russia's private sector, especially among small and medium-sized businesses and consumer industries. Most civilian sectors are experiencing economic contraction, evidenced by private sector activity reaching a three-year low in September 2025, with manufacturing and services both declining sharply while costs for suppliers, wages, and utilities rise faster than firms can offset through higher prices.[18]

Among Russia's twenty key manufacturing industries, only four are currently growing, and three of these are directly linked to military production. This reflects the broader shift of resources and investment toward the defense sector as sanctions and high interest rates stifle civilian industry profitability and growth.[19]

The auto industry illustrates these shifts: major manufacturers such as AvtoVAZ and Gorky Automobile Plant (GAZ) have moved employees to four-day work weeks due to plummeting sales, high interest rates, and diminished access to affordable credit. Employee incomes in some auto factories have dropped by about 20% since these measures were introduced, and factories are increasingly sending workers on forced leave to reduce costs and match declining demand. Simultaneously, investment in real estate has fallen sharply with various reports indicating a reduction of approximately 44% as developers cut back in response to tighter financing conditions and decreased market confidence.[20]

Inflation has become a persistent problem despite the Central Bank's efforts to contain it through high interest rates. Inflation stood at 8.1% in August 2025, with the Central Bank's benchmark interest rate at 17%. The Central Bank has pushed back its target of returning inflation to 4%, delaying it to mid-2026 as Kremlin spending priorities crowd out monetary policy objectives.[21]

Russia's public sector workers, including teachers, doctors, police, and pensioners, are among the hardest hit by the country's ongoing economic turmoil and rising inflation. Their incomes and benefits continue to be indexed to the official consumer inflation rate, which has lately hovered around 8–9% according to Russian government and international monitoring. However, the reality for many house-holds is far more severe: the actual increase in the cost of living, due particularly to food and essential services, often exceeds 20% per year.[22]

Wages and pensions for this segment are generally updated to match official inflation, but this adjustment fails to keep pace with real prices faced by ordinary Russians. Utilities, food, and transporta-tion have experienced spikes that outpace the reported inflation aver-age, resulting in an erosion of purchasing power for those relying on fixed or state-linked incomes. For example, services and food prices have surged above the headline rate, with services experiencing nearly 13% annual inflation and food close behind, far outstripping most income adjustments for these groups.[23]

This growing gap between the Kremlin's narrative of economic stabilization and the lived experience of much of the population has sharpened social discontent. Surveys and economic commentaries highlight a sense of alienation within Putin's core electorate who see promises of economic growth and stability contradicted by daily struggles with high prices and shrinking real incomes. Notably, pension-to-wage ratios are forecast to fall, worsening prospects for the elderly and further highlighting the regime's inability or unwillingness to cushion the most vulnerable.[24]

Perhaps most troubling for the Kremlin's long-term prospects, the Russian military-industrial complex has been fundamentally degraded by a decade of targeted international sanctions and the demands of the Ukraine war effort.[25] Russia's ability to produce military hardware has been severely impacted, and its capacity to innovate and adopt modern military technology has been constrained.

PUTIN'S LIMITED OPTIONS

Against this backdrop of military stalemate, economic strain, and dangerous provocations, Putin faces a menu of options but none of them palatable.

Option 1: Continue Escalation

Putin could double down on the current strategy, intensifying attacks on Ukrainian infrastructure while continuing to test NATO's resolve through airspace violations and other provocative actions. Putin has already warned Ukraine about striking near the Russian-occupied Zaporizhzhia nuclear plant and suggested Moscow could retaliate against Ukrainian nuclear facilities, stating ominously: "They still have functional nuclear power plants on their side. What prevents us from responding in kind? Let them think about this."[26]

However, this path carries enormous risks. NATO has demonstrated both capability and willingness to respond to airspace violations, as evidenced by the coordinated response to the Poland incident and the launch of Operation Eastern Sentry.

Further provocations could trigger a more robust NATO response, potentially drawing Russia into the direct conflict with the alliance that Putin has sought to avoid. Moreover, escalation does nothing to address Russia's fundamental problems which are the fuel crisis, economic stagnation, and unsustainable casualty rates.

Option 2: Negotiate from a Position of Weakness

Despite nine months of efforts by the United States to broker peace, including talks in Saudi Arabia, Oval Office meetings, and even a Trump-Putin summit in Alaska, there remains no end in sight to the conflict. Putin could accept negotiations but doing so from his current position would require significant concessions that would be ideologically and politically unacceptable.

Recent public opinion surveys indicate that war fatigue is increasing among Russians, with a record 66% of respondents favoring peace talks as of August 2025, according to the independent Levada Center. However, most Russians supporting negotiations do so only if they ensure that Russia retains its territorial gains and avoids outcomes perceived as a strategic defeat; large majorities remain opposed to returning captured territories under any peace agreement.[27] For Putin personally, accepting anything less than a clear victory would undermine the narrative that has sustained his rule and justified the enormous sacrifices demanded of the Russian people.

Option 3: Economic Adjustment and Austerity

Russia is attempting to stabilize its war-driven economy by enacting fiscal consolidation and economic restructuring, focusing heavily on raising revenue through increased taxation rather than deepening national debt or making broad spending cuts. In 2025, the government has introduced significant tax hikes on businesses and consumers including raising the VAT from 20% to 22%, lowering the small business VAT threshold, and introducing targeted taxes on gambling and "unhealthy" consumer goods to bolster revenue for ongoing defense expenditures.[28]

These tax measures are projected to bring in an additional 1.8 to 2.9 trillion rubles in 2026, providing substantial new resources for the national budget. However, according to the Russian Finance Ministry and supporting analyses, these revenues are being overwhelmingly directed toward financing defense and security spending accounting for about 41–43% of all government expenditure with little left for non-military sectors such as health, education, or social welfare. As a result, civilian programs face notable cutbacks, and the economic strains are felt most acutely outside the war-critical sectors.[29]

The fundamental problem is structural. The armed forces and military-industrial complex have become the main beneficiaries of the war economy and would be the primary losers from any significant decrease in military spending.[30] Any attempt to cut military expenditures would trigger personnel outflows and create barely manageable financial problems for state-owned defense corporations that have become dependent on government investments, subsidies, and loan guarantees. This would be compounded by a weak private sector that is in no condition to absorb displaced workers or take up economic slack.[31]

Option 4: Muddle Through
with the War of Attrition

The default option and the one Putin appears to be pursuing is to continue the grinding war of attrition, hoping that Ukraine's will to resist will eventually break before Russia's resources are exhausted. This strategy relies on several assumptions: that Russia's larger population and resource base will ultimately prevail; that Western support for Ukraine will wane; and that Russia can sustain current casualty and equipment loss rates for the foreseeable future.

Each of these assumptions is increasingly questionable. Ukraine has developed what one analyst calls "Europe's second-largest army, with almost one million men and women currently in uniform and a large reserve of battle-hardened combat veterans."[32] Ukrainian defense technology companies are focusing on domestic missile production, drone swarms, and AI technologies, maintaining a technological edge over Russian systems.[33]

Russia's economic capacity to sustain its attrition strategy appears increasingly doubtful. Defense spending currently accounts for approximately 6.3-8% of GDP in 2025, nearing levels not seen since the Cold War, with military expenditure projected around 15.5 trillion roubles ($145-160 billion) this year, an increase from previous years. Meanwhile, Russia's budget deficit has sharply widened; in the first seven months of 2025, it reached about 4.9 trillion roubles, over four times greater than in the same period of 2024 and reflecting a significant structural strain on government finance.[34]

Economist Vladislav Inozemtsev highlights the stark economic crossroads Russia faces: the government must either tighten monetary policy to curb inflation and risk recession or maintain the current monetary stance to support production but suffer escalating inflation. Neither path offers a viable long-term solution.[35] This combination of elevated defense spending and ballooning fiscal deficits raises serious questions about the sustainability of Russia's war-driven economy.

THE DILEMMA OF CHOICE

What makes Putin's position particularly precarious is that all of his options are constrained by a fundamental political-economic trap of his own making. The Russian political leadership has painted itself into a corner with a high and rising military budget that pushes it toward continued aggression even as the economic and human costs become unsustainable.

The military-industrial complex and armed forces have become powerful institutional actors and consolidated societal groups whose interests are now deeply intertwined with the continuation of the war. These groups would resist any moves toward demilitarization, creating internal political pressure for Putin to maintain current policies regardless of their long-term viability.

CONCLUSION

As of October 2025, Vladimir Putin faces the most consequential crisis of his presidency. Nearly four years into the Ukraine war, Russia has realized none of its initial objectives: it has failed to seize major new territory, failed to break Ukrainian resistance, and failed to fracture the Western alliance.

Instead, the campaign has brought catastrophic losses, sparked the worst fuel crisis in post-Soviet history, and driven NATO into its most substantial military buildup along Russia's borders in decades.

Putin's remaining choices range from grim to disastrous. Escalation risks direct conflict with NATO. Negotiating from a position of weakness would be a public concession of failure. Trying to adjust the economy without ending the war is arithmetically impossible. And persisting with a war of attrition is becoming untenable as Russia's military, economy, and society strain under a conflict with no clear path to victory.

His ideological commitment to the war, coupled with the structural limits of Russia's militarized economy, makes changing direction extraordinarily difficult. The country is trapped in a state of stagnation where it is unable to sustain the war indefinitely, yet equally unable to end it without risking economic and political collapse.

What began in February 2022 as a calculated bid to reclaim Russian power has devolved into a quagmire that endangers Putin's legacy and Russia's future. In October 2025, every Kremlin option amounts to a form of defeat with each option distinguished only by its timing, scale, and the degree of risk it poses to regional and global security.

CHAPTER 3

THE MOSAIC OF WAR: ENDGAME WITHOUT CLOSURE

B y the late fall of 2025, the war in Ukraine had transformed from a regional conflict into a sprawling global confrontation that defied conventional diplomatic resolution.

What had begun in February 2022 as Vladimir Putin's attempt to rapidly subjugate Ukraine had metastasized into a multidimensional struggle engaging the world's major powers, reconfiguring alliance structures, and testing the resilience of both the international economic order and the rules-based system that had emerged from the Cold War.

The conflict's fourth year revealed not an approaching conclusion but rather a complex mosaic of interlocking pressures, military, economic, political, and diplomatic, that made any path toward peace extraordinarily difficult to construct.

This chapter examines the late 2025 landscape as a moment when the war's full complexity became visible. Rather than moving toward resolution, the conflict had evolved into a condition that was simultaneously unsustainable and resistant to termination.

Understanding why is essential to grasping both the immediate challenges facing decision-makers and the longer-term questions about European security architecture that any serious peace framework would need to address.

THE DONBAS CRUCIBLE: PUTIN'S MINIMUM OBJECTIVE

By late 2025, Vladimir Putin had narrowed his strategic focus to what increasingly appeared as a minimum acceptable outcome: consolidating Russian control over the Donbas region.

The grandiose objectives that had animated the initial invasion, regime change in Kyiv, the subjugation of Ukraine as a buffer state, the demonstration of Russian military superiority, had been abandoned in all but the most propagandistic official rhetoric. What remained was a grinding campaign to capture territory that Putin could present to domestic audiences as justifying the enormous costs Russia had incurred.

The city of Petrovsk emerged as the symbolic and operational center of this reduced ambition. Its capture would allow Moscow to claim complete control of the Donetsk oblast, one of the two administrative regions comprising the Donbas. For Putin, this represented the bare minimum that might be packaged as victory, or at least as achievement sufficient to avoid the appearance of outright defeat.

The concentration of Russian offensive power on this objective was remarkable. Artillery barrages that had been distributed across multiple fronts now focused on a relatively narrow sector. Elite units that had been held in reserve or rotated through less active areas were committed to the Petrovsk axis. The Russian military command understood that failure here would not simply mean a tactical setback but would undermine the entire rationale Putin had constructed for continuing the war.

Ukrainian forces, for their part, had no illusions about Russian intentions. The concentration of Russian power created both danger and opportunity. Kyiv's military leadership recognized that a headlong Russian push toward Petrovsk, if met with effective Ukrainian resistance, could become a killing ground that would inflict casualties Moscow could ill afford.

The Ukrainian strategy was not to hold every meter of ground at any cost but rather to impose such a severe price on Russian advances that the broader Russian war machine would suffer damage from which it could not readily recover.

This approach reflected lessons learned across nearly four years of warfare. Ukrainian commanders had observed how Russian forces, when concentrated and committed to specific objectives, became vulnerable to systematic attrition. Mobile defense, coordinated artillery strikes, and the effective use of Western-supplied precision weapons allowed Ukrainian units to trade space for Russian losses when strategically appropriate. The goal was not to prevent any Russian advance but to ensure that each kilometer gained cost the Russian military forces and equipment it could not replace at a sustainable rate.

The fighting around Petrovsk thus became a test of wills and resources. For Russia, it was an attempt to demonstrate that concentrated military power could still achieve territorial objectives despite years of attrition. For Ukraine, it was an opportunity to prove that Russian offensive capacity had been degraded to the point where even narrowly focused campaigns would fail to deliver results proportionate to their costs.

The outcome of this struggle would shape both the military situation heading into potential armistice negotiations and the psychological calculations on both sides about what might be achieved through continued fighting.

ESCALATION BEYOND THE FRONT: INFRASTRUCTURE AS BATTLESPACE

While ground combat concentrated on the Donbas, both sides escalated their campaigns against each other's infrastructure. This dimension of the conflict had been present from the war's early stages but intensified dramatically in late 2025 as both Ukraine and

Russia sought to create strategic effects that could not be achieved through conventional military operations alone.

Ukraine's campaign of long-range strikes on Russian infrastructure represented a fundamental shift in the character of the war. Using a combination of Western-supplied weapons systems and indigenously developed capabilities, Ukrainian forces targeted oil refineries, storage facilities, transportation nodes, and command-and-control centers deep inside Russian territory.

These strikes served multiple purposes. Tactically, they complicated Russian logistics and reduced Moscow's ability to sustain operations in Ukraine. Strategically, they demonstrated that Russia's vast territory no longer provided meaningful protection from Ukrainian capabilities. Psychologically, they challenged the Russian government's narrative that the war remained confined to distant battlefields that posed no direct threat to Russian citizens.

The targeting of energy infrastructure proved particularly significant given its connection to the international sanctions regime that had evolved over the course of the war. By late 2025, Western sanctions had expanded beyond their initial focus on Russian state entities and oligarchs to encompass third countries purchasing Russian oil and gas. India and China, which had dramatically increased their imports of Russian energy at discounted prices, found themselves facing economic pressure from the United States and European Union.

This created a complex dynamic in which Ukrainian military strikes against Russian energy production facilities reinforced the economic isolation Moscow was experiencing through sanctions.

Russia's response was to intensify its own infrastructure campaign against Ukraine, particularly targeting the electrical grid and heating systems as winter approached. Moscow's calculation was brutally straightforward: if Ukrainian civilians faced a winter without reliable power and heat, domestic pressure on the Zelensky government to accept unfavorable peace terms would increase. Russian planners believed that popular resilience, which had proven remarkably strong through three previous winters of war, might finally crack under sustained attacks that threatened basic survival needs.

This infrastructure war created a devastating feedback loop. Ukrainian strikes on Russian facilities prompted Russian retaliation against Ukrainian infrastructure, which in turn justified expanded Ukrainian targeting of Russian strategic assets. The mutual escalation occurred even as both sides periodically expressed interest in negotiations, revealing how difficult it had become to establish any stable baseline from which diplomatic discussions could proceed.

INTERNAL PRESSURES: GOVERNANCE UNDER STRAIN

The late 2025 period exposed severe internal pressures within both Ukraine and Russia that complicated each government's ability to prosecute the war and pursue diplomatic solutions.

These domestic challenges, often under appreciated in conventional military analysis, proved as consequential as battlefield developments in shaping the conflict's trajectory.

In Ukraine, renewed allegations of corruption created a crisis of confidence precisely when maintaining Western support was most critical. Investigations revealed that reconstruction funds, military procurement budgets, and humanitarian assistance had been siphoned off through networks of officials operating at multiple levels of government.

The scale of the misappropriation shocked international observers and Ukrainian citizens alike. For a government that had built its wartime legitimacy partly on claims of fighting for a democratic, transparent European future, these revelations were devastating.

The timing could hardly have been worse. Western capitals, already facing domestic political pressures over the costs of supporting Ukraine, now confronted evidence that some portion of their assistance was being stolen. Congressional hearings in Washington and parliamentary inquiries in European capitals questioned

whether continued large-scale aid could be justified absent fundamental reforms in Ukrainian governance.

President Zelensky found himself in the paradoxical position of
needing to demonstrate anti-corruption credentials while simultaneously requiring the political flexibility to manage a wartime government where rigid accountability measures might impede operational
effectiveness.

This internal governance crisis affected Ukraine's negotiating
position in subtle but significant ways. Any peace framework would
require sustained Western security guarantees and reconstruction
assistance. The corruption scandals raised questions about whether
Western publics would support such long-term commitments.
Ukrainian officials recognized that resolving these governance issues
was not merely a matter of domestic politics but directly affected the
country's strategic position in potential peace negotiations.

Russia confronted its own mounting internal contradictions. The
economic impact of sanctions, combined with the costs of sustaining
military operations, had created a deepening crisis that Putin's
government struggled to manage.

The Russian economy had demonstrated remarkable resilience in
2022 and 2023, adapting to Western sanctions through import substitution, expanded trade with non-Western partners, and strict capital
controls. By late 2025, however, these adaptive mechanisms were
visibly failing. Consumer goods shortages, industrial production
declines, and the erosion of living standards affected ordinary
Russians in ways that propaganda could not entirely obscure.

Perhaps more significantly for regime stability, elite dissatisfac-
tion had grown as the war's costs became undeniable. The business
community, which had initially rallied around Putin or at least acqui-
esced to his war aims, increasingly viewed the conflict as an obstacle
to Russia's economic future. Security service professionals, long the
core of Putin's support base, privately questioned whether the military campaign served Russia's interests. Putin's response to these
pressures mixed repression with attempts to shift blame onto
external enemies and domestic scapegoats.

Yet even a authoritarian system capable of suppressing overt dissent faced challenges when the war's costs became impossible to ignore. The question confronting Moscow's leadership was not whether current conditions were sustainable, clearly they were not, but whether any conceivable peace settlement could be achieved that would preserve the regime's legitimacy and Putin's personal position.

THE DIPLOMATIC LABYRINTH: TOO MANY ACTORS, NO CLEAR FRAMEWORK

Throughout late 2025, various peace proposals and armistice frameworks circulated through diplomatic channels. What made the situation particularly complex was not the absence of ideas for ending the war but rather the proliferation of competing proposals advanced by different actors with divergent interests.

The sheer number of parties whose cooperation would be required for any sustainable settlement, Ukraine, Russia, NATO members, the European Union, and major non-Western powers including China, India, and Turkey, created a diplomatic labyrinth without clear pathways.

Several fundamental questions defined the diplomatic landscape.

First and foremost was the issue of what military forces would be positioned where in any post-conflict arrangement.

- Would Russian forces withdraw from all Ukrainian territory, or would some form of demilitarized buffer zone be established?
- What security guarantees would Ukraine receive, and from whom?
- Would NATO troops be stationed on Ukrainian soil, or would security arrangements take some alternative form?

These were not abstract questions but matters of immediate operational significance that would determine whether any ceasefire could be sustained.The territorial dimension proved equally vexing.

No serious observer believed Russia would voluntarily withdraw from all occupied Ukrainian territory, yet Ukraine's government remained committed, at least in public statements, to the restoration of its full territorial integrity including Crimea.

Various compromise proposals suggested intermediate arrangements, prolonged demilitarization, international peacekeeping forces, deferred status determinations, but each faced obstacles. Russia rejected any framework that might be interpreted as acknowledging the illegitimacy of its territorial claims. Ukraine could not accept any proposal that appeared to reward Russian aggression with permanent territorial gains.

Beyond these bilateral issues, the broader European security architecture required reconsideration. The war had demonstrated the inadequacy of pre-2022 arrangements and the need for a more robust deterrence posture along NATO's eastern flank. Poland, the Baltic states, and Romania had substantially enhanced their military capabilities and hosted rotating NATO deployments. Any peace framework would need to address Russian concerns about NATO's enhanced presence, concerns that Moscow itself had provoked through its invasion, while avoiding any arrangement that would compromise the alliance's ability to defend its members.

Non-Western powers added another layer of complexity. China had avoided taking clear sides, maintaining economic relationships with both Russia and the West while positioning itself as a potential mediator. India had increased its energy purchases from Russia while maintaining security cooperation with the United States through the Quad framework. Turkey had sold drones to Ukraine while keeping diplomatic channels open to Moscow.

Each of these powers had distinct interests that would need to be accommodated in any comprehensive peace framework, yet their preferences were often mutually incompatible.

The multiplication of diplomatic initiatives, proposals from the UN, from individual states, from regional organizations, paradoxically made peace more difficult to achieve. Each new proposal created additional reference points and set expectations that complicated subsequent negotiations. Parties could always point to some alternative framework that better served their interests, making it nearly impossible to build consensus around any single approach.

THE TRANSITION QUESTION: PUTIN AND THE FUTURE OF RUSSIA

Underlying all diplomatic discussions was a question that few official negotiations could explicitly address: what would happen to Vladimir Putin and the political system he had constructed?

Any serious peace framework needed to grapple with the reality that Putin had staked his regime's legitimacy on the war's outcome. A settlement that appeared to represent Russian defeat would threaten his political survival and potentially trigger instability within the Russian state.

This created a profound dilemma for Western policymakers and Ukrainian leaders. On one hand, many believed that a democratic Russia led by different leadership would be more likely to respect international norms and could be integrated into European security structures. On the other hand, actively working to undermine Putin's position risked provoking desperate responses from a nuclear-armed state led by a leader with nothing to lose.

The challenge was to create conditions under which a post-Putin political transition might occur without triggering the very instability and escalation that peace negotiations sought to avoid.

Various scenarios for Russian political evolution were discussed, though rarely in official forums. Some analysts suggested that sustained economic pressure combined with military stalemate might eventually prompt elite action to remove Putin. Others argued that only a clear military defeat would delegitimize the current leadership sufficiently to enable political change. Still others warned that any transition away from Putin could lead to even more nationalistic leadership rather than the liberal transformation Western observers hoped for.

The uncertainty surrounding Russia's political future meant that any peace framework would necessarily be provisional. Agreements negotiated with Putin's government might not survive a leadership transition, while waiting for such a transition risked indefinite continuation of the war.

This temporal dimension, the gap between immediate armistice arrangements and longer-term peace structures, reflected the deeper challenge of building stable security arrangements with a state whose internal evolution remained profoundly unpredictable.

NATO'S REBUILT FLANK:
THE NEW EASTERN EUROPE

While attention focused on the immediate battlefield and diplomatic initiatives, a quieter but equally significant transformation had occurred along NATO's eastern flank. The alliance members bordering Russia and Belarus had undertaken the most substantial military buildups since the Cold War, fundamentally altering the strategic landscape regardless of how the Ukraine conflict concluded.

Poland had emerged as a frontline military power, investing heavily in modern equipment including Abrams tanks, F-35 aircraft, and advanced air defense systems. The Polish government had also dramatically expanded its armed forces, viewing the war in Ukraine as validation of long-standing warnings about Russian intentions.

Similar buildups occurred in the Baltic states, Romania, and to a lesser extent in Slovakia and Hungary. These countries were not simply hosting NATO forces but developing substantial indigenous military capabilities.

This transformation created facts on the ground that would shape any post-war settlement. The notion that European security could return to pre-2022 arrangements was clearly impossible.

Russia now faced a far more capable NATO presence along its western borders, a development that Moscow would inevitably view as threatening but which alliance members considered essential defensive measures. Any peace framework would need to address this reality without compromising the defensive posture that had been established.

The consolidation of NATO's eastern presence also had implications for alliance unity and decision-making. Countries that had experienced Soviet occupation had different threat perceptions and risk tolerances than Western European states. This created potential fractures within the alliance, with eastern members generally favoring harder lines toward Russia than their western counterparts. Managing these internal differences while maintaining alliance cohesion would be crucial to any effective peace negotiation strategy.

ENDING THE CONFLICT IN UKRAINE

The fundamental asymmetry in this conflict is Russia's nuclear arsenal. This isn't merely about the abstract risk of nuclear war. It's about how nuclear weapons constrain the range of viable military strategies and political outcomes.

Western support for Ukraine has operated within carefully calibrated limits designed to help Ukraine defend itself without providing capabilities that could threaten Russian territory or regime stability. This explains the delayed and incremental provision of advanced weapons systems, restrictions on striking targets inside Russia, and general wariness about Ukrainian membership in NATO. The calculation, whether explicitly stated or not, is that pushing Russia toward complete military defeat carries unacceptable risks of nuclear escalation.

This creates a profound strategic problem. If the West cannot support Ukraine to the point of complete military victory due to nuclear risks, but also cannot accept Russian conquest of Ukraine, the only realistic path forward involves some form of negotiated settlement.

Yet any such settlement, by definition, means Russia retains at least some of its territorial gains, rewarding nuclear-backed aggression and potentially incentivizing future adventurism.

Critics of negotiated settlements argue that continued Western support can achieve better outcomes over time. This "long war" strategy rests on several assumptions:

- First, that sustained military aid can gradually degrade Russian capabilities to the point where Moscow concludes it cannot achieve its objectives militarily. The Ukrainian military has proven remarkably effective at destroying Russian equipment and personnel. Given time and continued support, this attrition might force Russia to accept terms closer to Ukraine's preferred outcome.
- Second, that economic sanctions will eventually create internal pressure within Russia for policy change. While sanctions haven't caused immediate economic collapse, they have cost Russia hundreds of billions in lost revenue and technological access. The cumulative effect over years or decades might prove decisive.
- Third, that Russian domestic politics might shift, through elite discontent, economic crisis, or even regime change, creating opportunities for a more favorable settlement. The Soviet Union's collapse, after all, came not from military defeat but internal systemic failure.

These arguments have merit, but they also carry significant costs and risks. The immediate cost is measured in Ukrainian lives and destruction. Every month the war continues means thousands more casualties and billions more in damage to Ukrainian infrastructure and cities. The psychological and demographic toll on Ukraine, with millions displaced and entire cities destroyed, grows more severe with time.

The long war also depends on sustained Western commitment, which is far from guaranteed. Political changes in the United States and Europe could dramatically reduce support levels. Western publics, facing their own economic challenges, may grow weary of the financial burden. Ukraine itself could exhaust its manpower and will to continue fighting at current intensity.

Perhaps most importantly, the long war carries its own escalation risks. A desperate or cornered Russia might resort to tactical nuclear weapons or strikes on NATO territory. The longer the conflict continues, the greater the cumulative probability of such catastrophic escalation.

What Would a Workable Settlement Look Like?

- Territorial arrangements that freeze current lines while explicitly leaving final sovereignty questions unresolved. This differs from recognizing Russian annexations. Think of it as a long-term ceasefire that acknowledges control without legitimizing it—similar to the Korean armistice that has held for over 70 years without resolving the underlying dispute.
- Multilateral security guarantees involving the United States, United Kingdom, France, and other major powers. These would need to be more substantial than vague promises, perhaps including forward-deployed forces, air defense systems, and explicit commitments to respond militarily to renewed Russian aggression. Short of NATO membership, this would need to be the most credible deterrent possible.
- Graduated sanctions relief tied to Russian compliance with specific, verifiable commitments. Rather than immediate lifting of all sanctions, create a framework where relief is earned through demonstrable good behavior over time.
- International monitoring of the ceasefire line and demilitarized zones, with mechanisms for verifying troop withdrawals and preventing military buildup.
- Reconstruction assistance focused on rebuilding Ukrainian infrastructure and economy, funded partially through controlled access to frozen Russian assets.
- Explicit provisions for renewed sanctions and military support if Russia violates the agreement. The settlement

must make clear that breaking the peace carries
immediate, severe consequences

The goal would be to create a situation where Ukraine can
rebuild and strengthen behind defensible lines, while Russia faces
sufficient deterrence against future aggression. It wouldn't satisfy
anyone's maximalist demands, but it might offer the best achievable
outcome given current realities.

The strongest argument against any territorial concessions is the
precedent it sets for international order. If nuclear-armed powers can
successfully conquer territory through aggression, what prevents
future attempts? China watching Taiwan, for instance, might
conclude that rapid conquest followed by negotiated retention of
gains is a viable strategy.

This concern is serious but perhaps overstated. The Ukraine war
has already established several important precedents regardless of
how it ends: that Western-backed resistance can impose massive
costs on aggressors, that economic sanctions can partially isolate
major powers, and that conquest of defended territory in the modern
era is extraordinarily difficult and costly. Russia has suffered
catastrophic military and economic damage for relatively modest
territorial gains. This is hardly an encouraging model for potential
aggressors.

Moreover, the precedent argument can cut both ways. If the West
commits to supporting Ukraine indefinitely in pursuit of complete
Russian withdrawal but then proves unable or unwilling to sustain
that commitment, the precedent becomes that Western security guarantees are unreliable. Better perhaps to negotiate the strongest
achievable settlement now than to promise support that ultimately
proves unsustainable.

Absent from much strategic analysis is the perspective of
Ukrainians actually living through this war. Every day brings drone
strikes, artillery bombardments, and the grief of lost loved ones.
Cities in eastern Ukraine have been reduced to rubble. Millions have

fled abroad, with uncertain prospects of return. The psychological trauma will reverberate for generations.

For these people, abstract debates about international precedent or long-term strategic competition may ring hollow against the immediate relief that peace would bring. Yet they also live with the knowledge that bad peace can simply be prelude to renewed war, that territorial concessions without credible security guarantees might only give Russia time to rebuild and attack again.

Ukrainian opinion on negotiations remains complex and divided. While most Ukrainians reject immediate territorial concessions, polling suggests growing openness to negotiated settlements as the war grinds on. The challenge for Ukrainian leadership is balancing the will to resist against the human cost of continued conflict—a calculation that becomes more difficult with each passing month.

The honest answer is that there are no good options, only choices between different sets of risks and costs. Continued war means sustained casualties and destruction with no guarantee of better eventual terms. Negotiated settlement means accepting that aggression has partially succeeded while hoping deterrence prevents future attacks. The long war might eventually produce internal Russian change, or it might simply exhaust Ukraine and Western support first.

What we cannot do is pretend these choices don't exist, or that righteous indignation about Russian aggression constitutes a strategy. The nuclear shadow over this conflict is real, and it fundamentally constrains what military outcomes are achievable without courting catastrophic risks. Within those constraints, the challenge is crafting a settlement that gives Ukraine the strongest possible security while creating meaningful deterrence against future Russian aggression.

The path forward likely requires accepting territorial ambiguity, frozen lines without recognized sovereignty changes, combined with the most robust security guarantees achievable short of NATO membership. It means using frozen Russian assets for Ukrainian reconstruction while maintaining sanctions leverage. It means building Ukraine's long-term defensive capabilities so that any future Russian attack would be prohibitively costly.

Will this satisfy demands for justice or complete territorial restoration? No.

But the alternative may be a grinding war of attrition that exhausts Ukraine without producing better outcomes, or a rushed settlement that leaves Ukraine vulnerable to renewed attack. Between these unappealing options, a carefully structured ceasefire with credible security guarantees might offer the best achievable outcome given the strategic realities we face.

The question is not whether we like these choices, but which uncomfortable reality we can best live with and which gives the Ukrainian people the greatest chance at security, reconstruction, and a future beyond war.

CONCLUSION: THE MOSAIC
REMAINS FRACTURED

By late fall 2025, the war in Ukraine had become a condition rather than an event or a complex system of military, economic, political, and diplomatic interactions that resisted resolution through conventional peace processes.

The concentration of Russian forces on Donbas objectives, the escalating infrastructure campaigns, the internal pressures within both warring states, the proliferation of incompatible diplomatic initiatives, the uncertainty about Russia's political future, and the transformation of NATO's eastern flank created a mosaic of challenges that no single agreement could address.

What made the situation particularly difficult was that the conflict had effectively globalized.

Ending the war was no longer simply a matter of negotiating terms between Kyiv and Moscow, or even between Russia and the West. The involvement of China, India, and other major powers as economic partners sustaining the Russian war effort, the implications for global energy markets and sanctions regimes, and the broader questions about international order all meant that any peace framework needed to operate at multiple scales simultaneously.

The war that had begun as Putin's attempt to restore Russian imperial control over Ukraine had evolved into something far more complex: a test of whether the post-Cold War international system could enforce basic norms of sovereignty and territorial integrity, a struggle over European security architecture, an economic contest between sanctions and adaptation, and a multidimensional competition engaging powers across the globe.

As 2025 drew toward its close, none of these dimensions had reached resolution. The mosaic of war remained visible in all its complexity, but the picture it formed offered no clear path toward peace. That challenge would define the months and years ahead.

SECTION VIII
CONCLUSION

History rarely repeats itself exactly, but it often rhymes with haunting resonance. The international coalitions that formed around the Spanish Civil War of 1936-1939 and the ongoing conflict in Ukraine since 2022 reveal striking parallels in how major powers align themselves during ideological and geopolitical crises. Both conflicts transformed from local disputes into global proxy wars that tested the resolve, capabilities, and alliances of the world's major powers. Yet the mechanisms of international involvement, the nature of the coalitions, and the broader strategic context have evolved dramatically between these two pivotal moments separated by nearly a century.

CHAPTER 1

HOW NUCLEAR WEAPONS
SHAPE THE WAR IN UKRAINE

Throughout the war in Ukraine, nuclear weapons have cast a long shadow over every strategic decision, every weapons shipment, and every escalatory step.

Yet Western discourse has largely treated nuclear threats as background noise or dangerous rhetoric from an increasingly desperate Putin rather than fundamental constraints shaping the entire character of the conflict.

According to Paul Bracken, Professor of Management and Political Science at Yale University and a leading expert on nuclear strategy, this represents a critical misunderstanding of how nuclear weapons function in 21st-century major power competition.

Bracken's central argument challenges conventional Western thinking: nuclear weapons are not simply sitting in silos waiting for an apocalyptic "use or lose" scenario. Instead, they are being actively employed right now to shape the strategic environment, create sanctuaries, deter interventions, and establish rules of engagement that favor nuclear-armed powers. In Bracken's blunt assessment when asked if Russia would actually use nuclear weapons: "My response was 'they already have.' It's a nuclear head game, and very dangerous."[1]

This perspective, that nuclear weapons are already in play rather than theoretically available for some future extremity, fundamentally reframes our understanding of the Ukraine conflict and its implications for global security. It suggests that the West is fighting a war constrained by nuclear dynamics while pretending those constraints don't exist, a dangerous form of strategic denial that echoes failures in Vietnam and Afghanistan.

THE LOST ART OF NUCLEAR CRISIS MANAGEMENT

To understand the current predicament, Bracken argues we must recognize what has been lost since the Cold War. During the 1960s through 1980s, the United States and Soviet Union developed sophisticated mechanisms for managing nuclear-shadowed crises.

These included:

- Robust back-channel communications that allowed leaders to signal intentions and test boundaries without public commitments that would lock them into dangerous positions. The Cuban Missile Crisis was resolved partly through such channels, with private assurances about Turkey offsetting public positions on Cuba.
- Professional communities of strategic thinkers outside government, figures like Herman Kahn and Thomas Schelling, who "broke the government's monopoly on strategic discourse and nuclear planning, to the benefit of much better thinking about what we were doing." These civilian strategists developed concepts like escalation dominance, intra-war deterrence, and limited nuclear options that gave policymakers frameworks for thinking about nuclear use beyond all-or-nothing scenarios.
- Negotiation architectures designed specifically for managing nuclear competition, from the Intermediate Nuclear Forces talks to Conventional Forces in Europe

negotiations. These weren't just arms control exercises; they were ongoing dialogues that created shared understandings of red lines, stabilizing measures, and acceptable behaviors.

- Alliance consensus-building processes that worked through contentious issues like the deployment of Pershing II and cruise missiles in Europe during the Euromissile crisis of the early 1980s. The United States and its European allies engaged in difficult, sustained negotiations to develop common positions before engaging the Soviets.

Today, virtually all of these capabilities have atrophied. As Bracken notes, "The kinds of capabilities you are referring to which existed in the 1980s are simply gone today."[2] The generation that navigated the near-nuclear war of 1983, when Soviet early warning systems detected a false American first strike during Able Archer exercises, has largely retired. Their skills in managing nuclear-tinged crises have not been passed down or maintained.

The consequences are visible in current policy debates, which Bracken describes as "a mish-mash, half-baked sets of ideas that pass for a strategic conversation in Washington today."[3] The Biden Administration focused on which specific weapons systems to provide Ukraine week by week, calibrating tactical actions without apparent strategic underpinning or long-term planning for conflict termination.

THE REALITY OF
NUCLEAR SANCTUARIES

Perhaps the most tangible impact of nuclear weapons on the Ukraine conflict is the creation of what amount to strategic sanctuaries or territories that neither side will attack despite their critical role in sustaining military operations.

On the Western side, Ukrainian forces train openly on British soil, learning to operate advanced systems like AS-90 self-propelled artillery. Similar training occurs across NATO countries, with Ukrainian troops cycling through Poland, Germany, and other allied states. Weapons flow through well-established supply chains from Western Europe into Ukraine. Intelligence fusion centers operate from NATO capitals. All of this happens in plain sight.

Russia tolerates this massive rear-area support operation not because it lacks the capability to strike these facilities, but because attacking NATO territory risks triggering Article 5 collective defense obligations and potentially escalating to nuclear exchange. The nuclear deterrent, British, French, and American, creates an inviolable sanctuary for Ukrainian force generation and sustainment.

The same dynamic operates in reverse. Russia supplies its operations in Ukraine from vast rear areas in Russian territory, staging troops, stockpiling ammunition, and operating command centers beyond Ukrainian reach. While Ukraine has struck targets in Russia with drones and special operations, these remain selective attacks rather than systematic campaigns against Russian infrastructure, logistics, and command facilities.

Western nations have explicitly limited the range of weapons provided to Ukraine to prevent strikes deep into Russian territory, most notably refusing long-range missiles that could threaten Moscow or major Russian military installations. This reflects fear that Russia would treat such attacks as existential threats potentially warranting nuclear response.

As Bracken noted in one of our discussions, this situation would be incomprehensible without nuclear weapons: "Does anyone really believe the Russians would tolerate this if not for British nuclear deterrence reinforced by the American nuclear force?" The answer is clearly no.

In any conventional-only conflict, both sides would systematically target each other's rear areas, supply lines, training facilities, and logistics networks. Nuclear weapons have created a different kind of war, one where the actual fighting is constrained to Ukrainian territory while the real sources of military power remain protected.

PUTIN'S NUCLEAR ALERT: DETERRENCE IN ACTION

The clearest demonstration of nuclear weapons' active role came in February 2022, when Putin ordered Russian nuclear forces to elevated alert status shortly after invading Ukraine. Western media largely treated this as bluster or a sign of Putin's supposed irrationality. Bracken sees something quite different: rational employment of nuclear signaling to achieve specific military and political objectives.

The purpose of the Russian nuclear alert, in Bracken's analysis, was twofold.

- First, it deterred NATO from massing conventional forces on the Belarus and Ukraine borders for potential intervention. In the early days of the war, when Russian forces seemed to be struggling badly, there was at least theoretical discussion of NATO establishing a no-fly zone or creating humanitarian corridors enforced by Western troops. The nuclear alert killed such discussions instantly.[4]
- Second, it signaled that any major NATO cyber or electronic warfare campaign to disrupt Russian air operations over Ukraine would be treated as an intolerable escalation. "A U.S. or NATO cyber campaign against distributed tactical nuclear and mobile missiles of Russia would manipulate the risk of escalation, which is why Putin ordered the alert."

This represents a sophisticated understanding of modern warfare. Russia knows its command and control systems, its tactical nuclear forces, and its air defense networks are potentially vulnerable to advanced Western cyber capabilities. By raising nuclear alert status, Russia warned that it would treat attacks on these systems, even non-kinetic attacks, as potential preludes to nuclear war, creating a deterrent barrier around not just its territory but its military capabilities.

The United States and NATO received the message. There has been no public evidence of major Western cyber campaigns against Russian strategic systems, despite clear Western advantages in this domain. Nuclear weapons created a "keep out" zone around certain categories of targets and operations.

THE WEAPONIZATION OF EVERYTHING

Bracken argues that we're witnessing a fundamental shift in how authoritarian powers think about warfare, with nuclear weapons enabling a broader strategy of comprehensive weaponization. China, Russia, and Iran "have moved to a position where weaponization of anything is an appropriate instrument of war, such as oil, food and certainly information."

This represents a departure from Western concepts that distinguish between military and non-military tools of statecraft, between legitimate targeting and civilian infrastructure, between soft and hard power. In the new paradigm, anything that can pressure adversaries becomes a potential weapon, and nuclear deterrence protects the state employing these hybrid strategies from conventional military responses.

Russia's cutting of the internet cable in Svalbard, Norway, in January 2022 exemplifies this approach. The cable serves SvalSat, the world's largest commercial ground station with over 100 satellite antennas providing critical communications for polar-orbiting satellites, including the International Space Station. Cutting it was a sub-threshold attack on space infrastructure, significant but ambiguous enough to avoid triggering collective defense.

Similarly, Russia weaponized energy supplies to Europe, using the threat of winter gas cutoffs to complicate European support for Ukraine. It weaponized food exports, threatening global grain supplies by blockading Ukrainian ports. It weaponized migration, deliberately creating humanitarian crises that send refugees toward European borders.

Nuclear weapons provide the umbrella under which these hybrid operations occur. Russia can employ aggressive gray-zone tactics while its nuclear arsenal prevents the kind of comprehensive conventional military response that might otherwise ensue. The West can impose economic sanctions but not the type of coercive military action that has historically been employed against non-nuclear powers.

THE MISSING STRATEGIC FRAMEWORK

What's notably absent from current discourse is serious consideration of how the war ends and what comes after. These questions require thinking beyond the weekly tactical cycle to consider fundamental issues:

- What forces remain and where after conflict termination?
- Do Russian forces stay in occupied Ukrainian territory?
- Does Ukraine accept the loss of Crimea and Donbass?
- What security guarantees does Ukraine receive from NATO, the United States from the European Union?

These questions determine whether we're creating a frozen conflict, a bitter peace that leads to renewed war, or some more stable arrangement.

What happens to Russia?

Some analysts advocate for Russian Federation disintegration as a desirable outcome. The dissolution of Russia would pose numerous problems, including the challenge of the disposition of thousands of nuclear weapons. The fall of the Russian Federation would also create mini states that could lead to Chinese domination of the Russian Far East and wars in Central Asia.

In fact we saw this before in the 1990s in the wake of the collapse of the Soviet Union. I was part of one team that worked this problem in detail for the U.S. government,

Yet there's little evidence this scenario is being seriously wargamed or that anyone has thought through how to manage such chaos.

Who pays for Ukrainian reconstruction?

The devastation is immense, requiring hundreds of billions in reconstruction. European states are already stretched thin, and American domestic politics makes massive Ukrainian reconstruction spending problematic. If the West doesn't step up, China likely will, fundamentally shifting Ukraine's orientation despite the war's outcome.

How is Russia reintegrated into the global economy?

Maintaining sanctions indefinitely against a nuclear-armed state of 140 million people creates permanent instability. "If we keep treating them as a pariah, and we maintain the sanctions after Putin is gone, then we are just asking for more of the same," as Pippa Malmgren noted in 2023.[5]

Yet there's no visible planning for how Russia might eventually rejoin international economic structures.

- What does this mean for NATO and European security?
- The war has already transformed NATO, adding Finland and Sweden, but what comes next?
- Do frontline states get extended-range strike weapons?
- How is forward defense postured?
- What role does nuclear deterrence play in European security?

These questions will define the next European security order, yet they're being answered by default through incremental decisions rather than strategic design.

LESSONS FROM THE 1980S

The contrast with how the Reagan Administration handled the Euromissile crisis is instructive. In the early 1980s, NATO faced a crisis over Soviet SS-20 intermediate-range ballistic missiles targeting Western Europe. The alliance response, deploying American Pershing II and ground-launched cruise missiles, was politically divisive, technically complex, and potentially destabilizing.

I spent a great deal of time in the 1980s in Europe working on these issues and much of that work was later expressed in my various books on nuclear weapons and European deterrence. There was a deep web of interaction among Western policy makers and analysts and with Soviet policy makers and analysts which facilitated to crisis management decision making.

But the process demonstrated capabilities that are missing today:

- Alliance consensus-building: The United States worked intensively with European allies to develop common positions before engaging the Soviets. This wasn't just consultation. It was genuine negotiation among allies about what risks they would accept and what objectives they shared.
- Strategic frameworks for negotiation: The West developed detailed positions on what it would and wouldn't accept in terms of missile deployments, verification regimes, and conventional force arrangements. These weren't worked out on the fly but resulted from sustained interagency and alliance coordination.
- Civilian strategic expertise: Figures outside government contributed critical thinking about nuclear strategy, deterrence stability, and crisis management. This created a rich debate that informed policy rather than simply supporting predetermined positions.
- Recognition of the end-to-end problem: Policy makers understood that deployment was not the goal. Stable

deterrence and eventually negotiated reductions were. Military capability was a means to diplomatic ends, not an end in itself.

The Intermediate Nuclear Forces Treaty that eventually resulted represented sophisticated arms control that addressed both sides' security concerns while reducing nuclear dangers. Nothing comparable seems possible today, partly because the institutional capabilities and expertise that made it possible have been allowed to atrophy.

IMPLICATIONS FOR FUTURE CONFLICTS

The Ukraine case creates precedents that will shape future conflicts, particularly regarding China and Taiwan. If nuclear weapons create inviolable sanctuaries even during active conflict, this fundamentally affects war planning throughout the Indo-Pacific.

- Would the United States be willing to strike targets in mainland China in a Taiwan conflict, knowing China's growing nuclear arsenal creates escalation risks?
- Conversely, would China refrain from striking U.S. bases in Japan, Guam, or even Hawaii due to American nuclear deterrence?
- How do conventional war plans account for nuclear constraints?
- What impact do the new technologies , AI and drones, on deterrence or the threat of accidental nuclear war?

The sanctuary dynamic also raises questions about concepts like Multi-Domain Operations that assume freedom of action across domains and geography. If nuclear weapons create de facto no-strike zones around homelands of nuclear powers, how does this affect strategies that depend on deep strikes and comprehensive targeting?

Furthermore, the Ukraine experience provides a powerful demonstration effect for potential nuclear aspirants. If possession of nuclear weapons allows you to invade neighbors while deterring intervention, or conversely to receive unlimited support without fear of direct attack, the incentive for proliferation increases dramatically.

As Paul Bracken notes in *The Second Nuclear Age: Strategy, Danger, and the New Power Politics* (2012), nuclear weapons provide states with geopolitical maneuver space, the freedom to act without fear of decisive conventional military response. This point underscores the logic behind Dmitry Medvedev's 2020s assertion that "the idea of punishing a country that has one of the largest nuclear potentials is absurd," illustrating the dynamic Bracken identified in the emergence of a second nuclear age.

Putin has demonstrated that nuclear weapons have many uses beyond the ultimate doomsday scenario. They create sanctuaries, deter intervention, enable hybrid warfare, and provide maneuver space for revisionist policies. As Bracken puts it, Putin "used nuclear weapons to deter NATO from forward conventional operations in Ukraine with their own forces."

The West's response cannot be to simply wish this reality away or treat nuclear threats as bluffs. It requires developing new frameworks for thinking about conflict in a multipolar nuclear world:

- Accepting that peer competitor conflicts occur under nuclear shadow, with this affecting everything from targeting to escalation management to conflict termination. This isn't defeatism; it's realism about the operational environment.
- Rebuilding capabilities for crisis management that have atrophied since the Cold War, including back-channels, professional communities of strategic thinkers, and institutional frameworks for managing nuclear-shadowed confrontations.
- Developing strategies for "orchestra conducting" rather than commanding, recognizing that in a multipolar world, the United States cannot simply impose its preferred

solutions but must coordinate with allies who have their own interests and constraints.

- Thinking seriously about end states and conflict termination rather than simply taking incremental steps and hoping acceptable outcomes emerge. This means difficult conversations about what outcomes are actually achievable and what price we're willing to pay for different objectives.

- Recognizing that competitive coexistence with nuclear-armed rivals may be the best achievable outcome rather than comprehensive victory or transformation of rival states. This was the essence of Cold War détente, and similar frameworks may be necessary today.

The war in Ukraine is not just about Ukraine. It's about establishing precedents and patterns for how major power competition occurs in the nuclear age. The specter of nuclear escalation isn't some abstract future concern. It's actively shaping the present conflict in ways the West has been reluctant to acknowledge. Until we recognize that nuclear weapons are already "in play," constraining options and creating new strategic realities, we cannot develop effective policies for managing this conflict or future ones.[6]

As Bracken noted in a discussion with me, "We are in early stages of learning to live in a multi-polar world not well described in terms projected from the American domestic political scene."

This quote from Paul Bracken highlights the significant shift in global dynamics toward a multipolar world, a system where multiple powerful states or centers of influence coexist, rather than a single superpower-dominated order. Bracken argues that the frameworks commonly used in American domestic political discourse do not adequately capture the complexities and changes happening at the international level.

Bracken's observation suggests that traditional American political narratives and strategic thinking, which often reflect the experience of unipolarity or bipolarity (such as during the Cold War), are increasingly out of touch with the new realities shaped by powers like China, India, Russia, and regional blocs. In a multipolar world, power is more distributed, interactions are less predictable, and U.S.-centric policy models or analogies fail to address emerging strategic risks and opportunities.

This transition can produce friction as old habits meet new realities, requiring greater adaptability and learning across defense, diplomacy, and economic statecraft. And this is clearly the impact of war in Ukraine going from a regionally contained conflict to becoming the global war in Ukraine.

CHAPTER 2
THE GLOBAL WAR IN UKRAINE AND GLOBALIZATION

The war in Ukraine, which began with Russia's full-scale invasion in February 2022, has accelerated profound shifts in the global economic system.

However, characterizing these changes as simply the end of globalization misses the complexity of what is actually occurring. Rather than witnessing de-globalization, we are experiencing a fundamental reconfiguration of global economic relationships, marked by the rise of economic security considerations, the fragmentation of supply chains, and the emergence of competing economic blocs.

This transformation represents not the death of international economic integration, but its evolution into a new, more politically conscious form.

THE PRE-WAR ECONOMIC ORDER

To understand the current transformation, we must first examine the economic architecture that preceded it. The three decades following the Cold War were characterized by hyper-globalization, a period of unprecedented economic integration driven by several interconnected factors.

The collapse of the Soviet Union eliminated the primary ideological and geopolitical barrier to global market integration. China's entry into the World Trade Organization in 2001 brought the world's most populous nation fully into the global trading system. Technological advances, particularly in communications and logistics, dramatically reduced the costs of coordinating complex international supply chains.

This period saw the emergence of genuinely global production networks. A single product might contain components manufactured in a dozen countries, assembled in another, and sold worldwide. The ideology of economic efficiency reigned supreme, with companies optimizing their operations to minimize costs without significant regard for geopolitical considerations. The mantra was just-in-time manufacturing, lean inventories, and the pursuit of comparative advantage wherever it could be found.

Energy markets epitomized this integration. European nations, particularly Germany, became heavily dependent on Russian natural gas, which was viewed primarily through an economic lens. Russian gas was abundant, relatively cheap, and delivered through established pipeline infrastructure.

The economic logic seemed unassailable, and political considerations were subordinated to market efficiency. This created a situation where Russia supplied approximately 40 percent of Europe's gas imports by 2021, with some countries like Germany relying on Russia for over half their gas supply.[1]

Similarly, Western economies became deeply intertwined with China not merely as a source of cheap consumer goods, but as an integral component of advanced manufacturing supply chains. China dominated the production of rare earth elements essential for modern electronics, controlled substantial portions of pharmaceutical ingredient manufacturing, and became the world's factory for everything from smartphones to solar panels.

THE VULNERABILITY REVEALED

The Ukraine war brutally exposed the strategic vulnerabilities inherent in this system. When Russia invaded Ukraine, Western nations faced an immediate dilemma. Imposing meaningful economic sanctions on Russia required accepting significant economic pain at home, particularly regarding energy supplies.

The conflict revealed that economic interdependence, rather than being purely a force for peace as many had argued, could become a weapon of coercion.

Europe's energy crisis in 2022 and 2023 demonstrated the real-world consequences of this vulnerability. As Russia reduced gas flows and eventually cut off supplies through major pipelines, European nations scrambled to find alternatives. Energy prices soared, industries faced potential shutdowns, and governments spent hundreds of billions on emergency measures to secure alternative supplies and cushion the economic impact on households and businesses. Germany, Europe's largest economy and most dependent on Russian energy, faced the prospect of deindustrialization as energy-intensive industries threatened to relocate to regions with more secure and affordable energy supplies.

The crisis also highlighted the interconnectedness of global systems in unexpected ways. Ukraine and Russia together accounted for a significant portion of global wheat exports, and the war's disruption of Ukrainian agriculture and Black Sea shipping routes triggered food price spikes that reverberated globally, contributing to inflation and food insecurity, particularly in developing nations dependent on grain imports.

FROM EFFICIENCY TO RESILIENCE

The primary shift catalyzed by the Ukraine war is a fundamental reorientation of economic thinking from pure efficiency toward resilience and security. This represents a paradigm change in how both governments and corporations approach economic decision-making. The principle of comparative advantage, which dominated economic thinking for decades, is being supplemented and sometimes superseded by considerations of strategic autonomy and supply chain security.

This shift manifests in several concrete ways.

- First, there is a renewed emphasis on redundancy and diversification in supply chains. Companies that previously optimized for just-in-time delivery with single-source suppliers are now building buffer stocks and cultivating multiple suppliers across different geographic regions. The goal is no longer purely minimizing costs, but rather ensuring continuity of operations even in the face of geopolitical disruption.
- Second, there is a growing prioritization of friend-shoring or ally-shoring, whereby nations seek to concentrate their most critical supply relationships within a circle of politically aligned partners. The United States and European Union are actively working to reduce dependence on China for critical technologies and materials, while simultaneously strengthening economic ties with each other and with partners like Japan, South Korea, and India. This does not mean complete decoupling, but rather a strategic reduction of dependencies in sectors deemed critical for national security or economic stability.
- Third, domestic industrial policy has returned with a vengeance after decades of market-oriented approaches. The United States' CHIPS and Science Act, which commits over $50 billion to rebuilding domestic semiconductor manufacturing capacity, exemplifies this trend. The European Union's various initiatives to boost domestic production of batteries, chips, and renewable energy technology follow similar logic. These policies represent a recognition that market forces alone may not

generate the level of domestic capacity deemed necessary for strategic security.

THE ENERGY TRANSFORMATION

Perhaps nowhere is the reconfiguration more dramatic than in energy markets. Europe's response to the cutoff of Russian gas has been nothing short of revolutionary. In less than two years, Europe has fundamentally restructured its energy supply system, a transformation that might have taken decades under normal circumstances.

The emergency response included rapidly expanding liquefied natural gas import capacity, with new terminals built at unprecedented speed. European nations signed long-term contracts with LNG suppliers in the United States, Qatar, and elsewhere. Germany, which had no LNG terminals before the war, constructed multiple facilities in months rather than years. This infrastructure investment represents a permanent shift in the geography of energy flows.[2]

Simultaneously, the energy crisis accelerated the push toward renewable energy. High fossil fuel prices made renewable energy more economically competitive, while energy security concerns provided additional policy motivation for reducing dependence on imported hydrocarbons. European investment in wind and solar power surged, and heat pump installations boomed as both households and governments sought to reduce gas consumption.

This energy transformation has geopolitical implications extending far beyond Europe. The reduced European demand for Russian hydrocarbons has forced Russia to seek alternative markets, primarily in Asia. However, the pipeline infrastructure that efficiently moved gas westward cannot easily be redirected eastward, and Russia has been forced to accept significant price discounts when selling to alternative buyers like China and India. This represents a substantial loss of revenue and geopolitical leverage for Russia.

Meanwhile, the United States has emerged as a major LNG exporter and energy power broker, strengthening its geopolitical influence. Middle Eastern gas producers have gained increased importance as reliable suppliers to Europe. The global energy map is being redrawn in ways that will have lasting consequences for international relations and economic power dynamics.

THE TECHNOLOGY DECOUPLING

The Ukraine war has accelerated an already emerging trend toward technological decoupling, particularly in the competition between the United States and China. While this process began before 2022, driven by concerns about intellectual property theft, national security implications of telecommunications infrastructure, and strategic competition, the war has intensified the urgency and expanded the scope of technological separation.

The United States has implemented increasingly stringent export controls on advanced semiconductors and semiconductor manufacturing equipment to China, explicitly aimed at preventing Chinese advances in areas like artificial intelligence and supercomputing that could have military applications. These controls represent a fundamental shift from the previous approach of allowing relatively free technology transfer in exchange for market access and economic benefits.

European nations, while historically more hesitant about technological decoupling from China, have also moved toward greater scrutiny of Chinese involvement in critical infrastructure. The removal of Huawei equipment from 5G networks across multiple European countries, restrictions on Chinese investment in sensitive technologies, and increased vetting of research partnerships all reflect growing concern about technological dependencies.

This technological separation is creating parallel technology ecosystems, with different standards, protocols, and supply chains. The implications are profound. Companies face the prospect of having to maintain separate operations for different markets. Research collaboration becomes more difficult when security concerns restrict information sharing. The global innovation ecosystem, which thrived on relatively free exchange of ideas and talent, faces new barriers.

THE EMERGING BLOC STRUCTURE

Rather than a simple bifurcation into Western and non-Western blocs, the emerging global economic structure is more complex and fluid. Several distinct groupings are taking shape, with nations navigating between them based on their specific interests and constraints.

The United States and its close allies, including the European Union, Japan, South Korea, Canada, and Australia, form a core bloc characterized by democratic governance, market economies, and security alignment. This group is working to strengthen internal economic ties while reducing critical dependencies on nations outside the bloc.

China leads a second major grouping, seeking to develop its own sphere of economic influence through initiatives like the Belt and Road program and by positioning itself as a champion of Global South interests. However, China faces challenges in this role, as many developing nations are wary of excessive dependence on any single major power, particularly after witnessing Europe's difficulties with Russian energy dependence.

Perhaps most interesting is the emergence of a group of swing states that are actively maintaining relationships with multiple blocs while avoiding exclusive alignment with any. India epitomizes this approach, strengthening security and economic ties with the United States and its allies through frameworks like the Quad, while simultaneously maintaining substantial economic relationships with Russia and China. Many Southeast Asian, Latin American, and African nations are pursuing similar strategies of strategic autonomy.

This multipolar structure is more fluid and complex than the Cold War's rigid bipolar division. Trade and investment continue across bloc boundaries, but with more careful attention to strategic implications. The result is not complete separation, but rather a selective decoupling in sensitive sectors while maintaining integration in others.

ECONOMIC WARFARE AND FINANCIAL WEAPONIZATION

The Ukraine war has witnessed an unprecedented escalation in the use of economic tools as instruments of geopolitical competition. The Western sanctions imposed on Russia represent the most comprehensive economic warfare campaign ever directed at a major economy. Beyond trade restrictions, these measures included freezing Russian central bank reserves held abroad, excluding Russian banks from the SWIFT international payment system, and implementing secondary sanctions threatening penalties for third parties that help Russia evade restrictions.

The effectiveness and limitations of these measures have provided important lessons. The sanctions have significantly damaged the Russian economy, cutting off access to advanced technology, limiting ability to fund military operations, and constraining economic growth. However, they have not achieved their maximal objective of forcing a Russian withdrawal from Ukraine, and Russia has demonstrated considerable adaptability in finding workarounds through third countries and parallel systems.

The use of financial systems as weapons has generated unintended consequences. Many nations, including major U.S. partners, have begun questioning their vulnerability to similar measures. This has accelerated interest in alternative payment systems and reduced dollar dependency. China has intensified efforts to internationalize its currency and develop alternatives to dollar-denominated trade. Central banks worldwide have increased gold purchases as a hedge against financial system vulnerabilities.

These developments represent a potential long-term challenge to the dollar's dominant role in international finance, though such transitions typically occur over decades rather than years. The immediate effect is to make the international financial system more fragmented and less efficient, with multiple parallel systems operating simultaneously.

WINNERS AND LOSERS
IN THE NEW ORDER

The reconfiguration of global economic relationships is creating distinct winners and losers, though the picture remains fluid as the new system takes shape.

Among the clear beneficiaries are nations positioned to serve as alternative suppliers or intermediaries in the new system. Vietnam, Mexico, and Poland have attracted significant manufacturing investment as companies diversify away from China. Energy exporters like the United States, Qatar, and Norway have gained from redirected energy flows and higher prices. India has benefited from its ability to purchase discounted Russian oil while maintaining good relations with Western nations.

The losers include nations heavily dependent on the old system of free-flowing, purely efficiency-driven globalization. Russia's economy has been significantly damaged by sanctions and the loss of its primary export markets. Chinese export sectors face challenges from Western efforts to reduce dependence. Developing nations dependent on stable, cheap food and energy imports have suffered from market disruptions and price volatility.

For Western economies, the picture is mixed. They have successfully reduced critical vulnerabilities, particularly in energy, but at substantial cost. The transition has driven inflation, reduced living standards, and required massive government spending. Industrial competitiveness has been challenged by higher energy costs. The long-term outcome will depend on whether investments in resilience and domestic capacity generate sustainable economic advantages.

WHAT THIS MEANS FOR THE FUTURE

The transformation catalyzed by the Ukraine war is likely to be enduring, but its ultimate form remains uncertain. We are not witnessing the end of globalization, but rather its evolution into a more politically conscious, security-oriented form.

Several key trends seem likely to persist.

- First, economic efficiency will no longer be the sole or even primary consideration in critical sectors. Resilience, security, and political alignment will weigh heavily in decisions about supply chains, investment, and technology sharing. This will likely mean somewhat higher costs and reduced economic efficiency compared to the pre-war system, though proponents argue these are reasonable prices for reduced vulnerability.
- Second, government involvement in economic decision-making will remain elevated compared to the market-dominated approach of recent decades. Industrial policy, strategic investment, and active management of international economic relationships will be normal features of governance across developed economies.
- Third, the global economy will operate with more friction and fragmentation than in the past three decades. Multiple parallel systems for payments, technology standards, and supply chains will coexist. Cross-bloc economic activity will continue but with more scrutiny and restriction in sensitive areas.
- Fourth, the geopolitical competition between major powers will increasingly play out through economic means. Trade policy, investment restrictions, technology controls, and financial sanctions will be prominent tools of statecraft. This economic dimension of great power competition will likely intensify before any stabilization occurs.

CONCLUSION

The war in Ukraine has indeed marked a watershed moment in global economic history, but not simply by ending globalization. Instead, it has accelerated a fundamental reconfiguration of the international economic system from one organized primarily around efficiency and profit maximization toward one that explicitly incorporates security, resilience, and geopolitical considerations.

This transformation involves real costs. The new system will likely be less efficient, more expensive, and more complex to navigate than its predecessor. However, the old system's vulnerabilities and the coercive potential of economic interdependence justified this change. The challenge now is managing this transition while minimizing economic disruption and avoiding a slide toward complete fragmentation that would sacrifice the substantial benefits of international economic cooperation.

What emerges will not be a return to the pre-globalization era, nor will it be a complete severing of international economic ties. Instead, we are likely to see a more complex, multi-bloc global economy where international economic engagement continues but within a framework more explicitly shaped by geopolitical and security considerations.

The full contours of this new system will take years to crystallize, but the direction of travel is now clear. The era of purely market-driven hyper-globalization has ended, replaced by a more politicized, security-conscious form of international economic integration.

Success in this new environment will require adapting to a world where economic and security considerations are inextricably linked. The simplifying assumptions of the past, that economic interdependence automatically promotes peace or that markets can be separated from politics, no longer hold. The task ahead is navigating this more complex reality while preserving the benefits of international cooperation and avoiding the worst outcomes of economic nationalism and fragmentation.

CHAPTER 3

THE GLOBAL DIFFUSION OF DRONE WARFARE SKILLS

T he war in Ukraine has emerged as an unprecedented proving ground for modern warfare technologies, particularly in the realm of unmanned aerial systems.

Beyond its immediate European theater, this conflict has generated global security implications through the recruitment of foreign fighters who are acquiring highly specialized combat skills, especially in drone warfare, that may eventually migrate to conflicts and criminal enterprises in their home regions.

This chapter examines the documented evidence of foreign fighter participation in Ukraine, with particular focus on African fighters recruited by Russia and Colombian veterans fighting for Ukraine, and analyzes the growing concern among security experts that the tactical knowledge these fighters are acquiring, particularly in first-person view (FPV) drone operations and improvised munitions, represents a qualitatively new and potentially destabilizing form of technological transfer.

AFRICAN FIGHTERS
IN RUSSIAN SERVICE

Multiple credible sources have documented Russia's systematic recruitment of African fighters to supplement its forces in Ukraine, often through deceptive or coercive means. In November 2025, Ukrainian Foreign Minister Andrii Sybiha publicly stated that at least 1,436 Africans from 36 countries were serving with Russian forces. This remarkable statistic, derived from Ukrainian intelligence assessments and prisoner interrogations, reveals the geographic breadth of Russia's recruitment network, which has extended to Uganda, Kenya, Togo, Rwanda, Burundi, Congo, South Africa, and numerous other nations across the continent.[1]

The mechanics of this recruitment are particularly troubling. Rather than straightforward military contracts, Russian recruiters have employed a variety of deceptive tactics. PBS NewsHour investigations documented cases where African migrants already in Russia were essentially coerced into signing military contracts under threat of deportation, while others were lured with false promises of non-combat jobs, educational opportunities, or lucrative employment.[2]

The recruitment pipeline has become increasingly sophisticated, with Ukraine's Defence Intelligence (HUR) reporting in May 2024 that Russia was offering sign-up bonuses of $2,000, monthly pay of $2,200, and promises of Russian passports, not just for the recruits themselves but for their families as well.[3]

Once enlisted, these African fighters have faced particularly dire circumstances on the battlefield. Ukrainian officials and captured prisoners consistently describe how foreign fighters, especially Africans, are disproportionately assigned to what Russian forces term 'meat assaults' or human wave attacks against heavily fortified Ukrainian positions with minimal preparation or support. These tactics result in extraordinarily high casualty rates.

African Defense Forum reporting highlighted testimony from captured fighters who described being forced into suicidal attacks despite visible enemy armor, artillery, and drone coverage.[4] The expendability of these foreign fighters is further underscored by the fact that, according to Ukrainian POW officials, not a single African prisoner has been exchanged over the past year, with Russian authorities showing no interest in requesting their return.[5]

The human dimension of this recruitment is particularly stark in individual testimonies. A captured Somali fighter told journalists he had joined the Russian army to secure a 'good future' for his family and was promised he would serve as a 'helper' performing logistics and first aid duties.

Instead, he was immediately deployed to frontline assault positions without language training or adequate preparation. Similarly, a Sierra Leonean man recounted through tears how he paid a recruiter and flew to Russia for what he believed was legitimate employment, only to realize after signing Russian-language documents that he had inadvertently enlisted in the military. These cases reveal a pattern not of mercenary activity but of systematic exploitation of economic desperation in poor African communities.

COLOMBIAN VETERANS IN UKRAINIAN SERVICE

While Russia's recruitment of Africans has relied heavily on deception and coercion, Ukraine's enlistment of Colombian veterans represents a different model, one of professional military volunteers seeking both economic opportunity and a return to combat roles. Multiple sources confirm that hundreds, and potentially thousands, of Colombian former soldiers have enlisted in Ukrainian forces, primarily through the International Legion of Territorial Defense of Ukraine but increasingly within regular Ukrainian brigades.

The scale of Colombian participation is substantial. Euromaidan Press reporting from October 2024 quoted Colombian volunteers estimating between 1,000 and 2,000 of their countrymen were serving in Ukraine.[6]

While exact figures remain classified for security reasons, the presence of entire Spanish-speaking units and the establishment of

Spanish-language recruitment infrastructure by Ukraine's International Legion confirms significant Colombian involvement. These fighters bring substantial military experience from Colombia's decades-long internal conflict against guerrilla groups and drug cartels, having developed expertise in jungle warfare, counterinsurgency operations, and small-unit tactics.

The motivations driving Colombian veterans to Ukraine are complex but primarily economic. Former Colombian soldiers interviewed by *The World* described receiving monthly pensions of just $300 to $600 after retirement, barely sufficient to support their families or service accumulated debts.[7]

In contrast, Ukrainian forces offer frontline soldiers approximately $3,000 to $4,800 per month depending on deployment status, a transformative sum for families facing economic hardship in Colombia.

However, economic necessity is not the sole factor. Multiple Colombian volunteers described a sense of frustration with civilian life, a desire to return to military service where they felt purpose and belonging, and in some cases genuine ideological commitment to defending Ukraine against Russian aggression.

Individual profiles illuminate the human dimension behind these statistics. Edwin, a Colombian veteran with the call sign 'Thanos,' served ten years in Colombia's border guard forces, earning combat decorations including the Colombian Struggle Medal and the Sword of Freedom. When Russia launched its full-scale invasion, he resigned his military post and volunteered for Ukraine's International Legion, seeing it as an opportunity to help defend freedom in a nation he had never visited. His experience reflects a broader pattern among Colombian volunteers who, despite significant combat experience against guerrillas and cartels, have found Ukraine's high-intensity conventional warfare profoundly different and more dangerous than their previous service.[8]

Colombia's government has taken a nuanced position on this phenomenon. While not officially endorsing foreign military service, Colombian authorities have been reluctant to criminalize veterans who volunteer for Ukraine, recognizing both the economic pressures driving the migration and the legitimate nature of Ukraine's defensive war.

However, the Colombian military's Association of Retired Soldiers has expressed concern about the trend, particularly after incidents where Colombian volunteers were detained by Venezuela and transferred to Russia, where they face lengthy prison sentences on mercenary charges.[9] Ukraine has consistently maintained that these fighters are not mercenaries but volunteers integrated into its regular armed forces, a distinction with significant implications under international humanitarian law.

THE DRONE WARFARE
SKILLS TRANSFER CONCERN

The most significant long-term security implication of foreign fighter participation in Ukraine lies not in the immediate combat effects but in the specialized skills these individuals are acquiring—particularly in the operation, construction, and tactical employment of FPV drones and the integration of commercial unmanned systems with improvised munitions.

A March 2025 analysis by the Combating Terrorism Center at West Point explicitly warned that foreign fighters in Ukraine are becoming a cohort of 'highly lethal drone operators' whose expertise could be repurposed for terrorism, organized crime, or future insurgencies.[10]

Ukraine has emerged as the world's foremost laboratory for small unmanned aerial system warfare. By early 2025, Ukraine had established approximately 500 drone manufacturers and was producing an estimated 4 million drones annually—a scale unprecedented in modern warfare. The Ukrainian military created an entirely new branch of service dedicated exclusively to drone operations, with FPV drones reportedly responsible for up to 70-80 percent of Russian battlefield casualties.[11]

These first-person view racing drones, typically costing between

$200 and $1,000, have been adapted to carry various payloads including RPG warheads, thermobaric rounds, and improvised explosives, and are deployed in swarm tactics that overwhelm traditional air defenses.

The technological sophistication of drone warfare in Ukraine has accelerated dramatically. Both Russian and Ukrainian forces have developed fiber-optic controlled drones immune to electronic jamming, AI-assisted targeting systems, thermal imaging capabilities for night operations, and interceptor drones designed to destroy enemy UAVs in flight.

The West Point analysis emphasized that this represents a qualitatively different skill set from what previous foreign fighter cohorts acquired in conflicts like Afghanistan or Syria. The combination of FPV piloting, 3D-printed munitions fabrication, improvised payload integration, electronic warfare understanding, and swarm coordination tactics creates an operator with capabilities that could be extraordinarily dangerous if applied in terrorist or criminal contexts.

Evidence of this skills transfer is already emerging. Defense News reported in November 2025 that members of Latin American drug cartels had joined Ukraine's foreign fighter volunteer units specifically to gain FPV drone training, with subsequent deployment of these tactics along the U.S.-Mexico border.[12]

Mexican authorities documented over 260 explosive-equipped drone attacks by cartels in 2023 alone, representing a dramatic escalation in sophistication from earlier rudimentary attempts. U.S. Northern Command reported over 1,000 drone intrusions across the U.S.-Mexico border monthly in 2024, prompting the formation of specialized counter-drone rapid response teams to protect American military installations.

The Atlantic Council's research on this phenomenon argues that the barrier to entry for sophisticated drone operations has collapsed, meaning capabilities once exclusive to nation-states are now accessible to non-state actors including criminal syndicates and terrorist organizations.[13]

West Point analysts note that violent extremist organizations will likely adopt and adapt innovations emerging from Ukraine, including swarm tactics, advanced counter-drone measures, and AI-assisted targeting, creating what they describe as 'a new era of asymmetric warfare.'

The psychological impact of drone attacks, their precision, unpredictability, and the sense of omnipresent surveillance they create, makes them particularly attractive to groups seeking to generate terror effects with limited resources.

POLICY IMPLICATIONS AND THE ABSENCE OF REINTEGRATION ARCHITECTURE

Despite the clear risks identified by counterterrorism experts, current policy responses remain inadequate. The West Point analysis explicitly states that 'no serious reintegration architecture yet exists to manage this risk', a striking admission given the scale of foreign fighter participation in Ukraine.

Unlike previous foreign fighter cohorts, where radicalization and ideological extremism were primary concerns, the Ukraine cohort presents different challenges: they are acquiring highly technical, militarily valuable skills in a conflict widely viewed as legitimate self-defense, making it difficult to criminalize their participation or justify intrusive monitoring upon return.

The African fighters present particular policy dilemmas. Many have been effectively abandoned by both Russia and their home governments, with no exchange mechanism and the prospect of criminal prosecution should they return home. This creates perverse incentives for continued military service and potential recruitment by other actors offering employment for their combat skills.

Colombian veterans face different but equally complex challenges. While not subject to the same coercive recruitment as Africans, they too lack structured transition programs. Some Colombian training centers have begun offering tactical medicine and combat courses explicitly geared toward preparing veterans for foreign military service, creating an informal privatization of military expertise that operates in legal grey areas.

International humanitarian law experts note that the distinction between lawful volunteer service in regular armed forces and unlawful mercenary activity remains contested, particularly when economic motivation is primary. This legal ambiguity complicates efforts to regulate or monitor the phenomenon.

Meanwhile, the technological genie is out of the bottle. NATO members including Britain and Denmark are reportedly receiving training in drone warfare from Ukrainian military instructors, legitimizing and accelerating the diffusion of these capabilities. As Atlantic Council researchers observe, Ukraine has become a 'drone superpower' whose tactical innovations are being studied and adopted by militaries worldwide—both state and non-state.[14]

CONCLUSION

Open-source evidence now firmly establishes three interrelated phenomena: first, Russia's systematic recruitment of African fighters through deceptive means, often consigning them to the highest-casualty roles; second, Ukraine's successful integration of Colombian and other Latin American veterans as professional volunteers within its armed forces; and third, the transformation of Ukraine into a global training ground for advanced small-drone warfare tactics whose implications extend far beyond the current conflict.

The drone warfare skills being acquired by foreign fighters, from FPV piloting to improvised munitions integration to electronic warfare techniques, represent a qualitatively new form of combat expertise that differs substantially from capabilities developed in previous conflicts.

This technological transfer occurs in a vacuum of policy response. No comprehensive international framework exists for tracking, monitoring, or facilitating reintegration of foreign fighters with advanced technical military skills. The legitimate nature of Ukraine's defensive war and the economic desperation driving much recruitment complicate efforts to criminalize or stigmatize participation.

Meanwhile, early evidence from Mexico and elsewhere suggests that drone warfare techniques learned in Ukraine are already being adapted for use by criminal organizations and potentially terrorist groups. As the war continues and more foreign fighters rotate through Ukrainian training programs and combat operations, the global diffusion of these capabilities will accelerate, creating security challenges that will persist long after the conflict in Ukraine concludes.

Addressing this emerging threat will require unprecedented international cooperation in areas including technology export controls, foreign fighter monitoring, veteran reintegration programs, and counter-drone capability development—policy infrastructure that currently does not exist at the necessary scale or sophistication.

CHAPTER 4
ECHOES OF THE 1930S

History rarely repeats itself exactly, but it often rhymes with haunting resonance. The international coalitions that formed around the Spanish Civil War of 1936-1939 and the ongoing conflict in Ukraine since 2022 reveal striking parallels in how major powers align themselves during ideological and geopolitical crises.

Both conflicts transformed from local disputes into global proxy wars that tested the resolve, capabilities, and alliances of the world's major powers. Yet the mechanisms of international involvement, the nature of the coalitions, and the broader strategic context have evolved dramatically between these two pivotal moments separated by nearly a century.

THE SPANISH LABORATORY:
PRELUDE TO GLOBAL WAR

The Spanish Civil War emerged from a perfect storm of domestic political polarization and international tension. When Francisco Franco's military uprising began in July 1936, Spain quickly became the dress rehearsal for World War II. The conflict drew in major European powers not merely as observers or diplomatic mediators, but as active participants testing their military capabilities and ideological commitments.

The international coalitions that formed around Spain reflected the broader political alignments of the 1930s. On one side stood the fascist powers: Nazi Germany and Fascist Italy threw their support behind Franco's Nationalist forces. Hitler's Germany provided the infamous Condor Legion, an expeditionary force that gave German pilots combat experience and allowed testing of new aircraft and bombing tactics. The bombing of Guernica in April 1937 became a symbol of fascist brutality but also demonstrated Germany's willingness to experiment with terror bombing against civilian populations.

Mussolini's Italy committed even more extensively, sending approximately 50,000 troops, modern aircraft, and substantial military equipment to support Franco. For Mussolini, Spain represented an opportunity to challenge French and British influence in the Mediterranean while advancing fascist ideology.

Opposing them, the Soviet Union under Stalin became the primary international supporter of the Spanish Republic. Moscow provided military advisers, pilots, tanks, and aircraft, though this support came with significant political strings attached. Soviet involvement allowed Stalin to purge Spanish communists and anarchists who didn't align with his vision, extending his domestic terror campaigns into the international arena.

The democratic powers, Britain and France, officially maintained neutrality through the Non-Intervention Committee, a diplomatic fiction that allowed fascist powers to intervene while constraining aid to the Republic. This neutrality reflected both domestic political divisions and strategic miscalculations about the nature of the fascist threat.

MODERN PROXY WAR: THE
UKRAINE COALITION STRUCTURE

The Russian invasion of Ukraine in February 2022 triggered the formation of international coalitions that echo but don't precisely mirror those of the Spanish Civil War. The supporting coalition for Ukraine centers on NATO and EU member states, with the United States providing the largest share of military and financial assistance. This Western coalition has demonstrated remarkable unity, coordinating sanctions regimes, military aid packages, and diplomatic isolation of Russia with a level of institutional sophistication unavailable in the 1930s.

The breadth of support for Ukraine extends beyond traditional Western allies. Countries like Japan, South Korea, and Australia have provided significant assistance, reflecting how the conflict has been framed as a defense of the international rules-based order rather than merely a European regional dispute.

Russia's coalition presents a more complex picture. While China provides crucial economic and diplomatic support, it has been careful to avoid direct military assistance that might trigger Western sanctions. Iran has supplied drones and military technology, while North Korea has provided artillery shells and missiles. This coalition reflects shared antagonism toward Western hegemony rather than ideological alignment, marking a significant difference from the more ideologically coherent fascist alliance of the 1930s.

MECHANISMS OF
INTERNATIONAL INVOLVEMENT

The methods by which international powers have involved themselves in these conflicts reveal both continuities and dramatic changes in the nature of modern warfare and diplomacy.

In Spain, international involvement was direct and relatively transparent despite official denials. German and Italian pilots flew combat missions, Soviet advisers directed military operations, and foreign volunteers formed entire military units like the International Brigades. The technology of the 1930s required physical presence for effective military assistance for pilots had to fly the planes, advisers had to be present to operate complex equipment, and meaningful intelligence sharing required face-to-face coordination.

The Ukraine conflict operates in a fundamentally different technological environment. Satellite intelligence can be shared in real-time across continents, precision-guided munitions can be operated with minimal training, and cyber warfare capabilities can be deployed without any physical presence. This has enabled high-tech proxy warfare or the provision of sophisticated military capabilities without the direct personnel commitments that characterized earlier conflicts.

Western support for Ukraine has emphasized providing advanced defensive and offensive systems from Javelin anti-tank missiles to HIMARS rocket systems to sophisticated air defense networks while maintaining the fiction of non-belligerent status. Training of Ukrainian forces occurs outside Ukraine's borders, intelligence is shared through secure channels, and even targeting information is provided for strikes deep into Russian territory.

Russia's supporters have adopted different approaches reflecting their capabilities and constraints. Iran's provision of Shahed drones represents a new model of asymmetric military assistance, relatively inexpensive systems that can be produced at scale and cause disproportionate impact. North Korea's supply of artillery ammunition demonstrates how isolated states can still contribute meaningfully to modern conflicts through traditional military hardware.

IDEOLOGICAL FRAMING
AND GLOBAL NARRATIVES

Both conflicts became global ideological battlegrounds, but the nature of the competing ideologies and their appeal has evolved significantly.

The Spanish Civil War was framed in stark terms that resonated globally: fascism versus democracy, capitalism versus socialism, tradition versus modernity. These ideological frameworks had clear adherents worldwide and inspired volunteers to travel thousands of miles to fight for their beliefs. The International Brigades attracted approximately 35,000 volunteers from over 50 countries, including prominent writers like George Orwell and Ernest Hemingway who chronicled the conflict for global audiences.

The ideological framing of the Ukraine conflict is both more complex and more contested. Western narratives emphasize democracy versus authoritarianism, international law versus aggression, and the rules-based order versus revisionist powers. These frames resonate strongly in liberal democracies but have less universal appeal than the anti-fascist narrative of the 1930s.

Russia's counter-narrative focuses on resistance to Western hegemony, protection of traditional values against liberal decadence, and the defense of a multipolar world order. This messaging appeals to countries and populations that have experienced Western intervention or feel marginalized by the current international system, though it lacks the revolutionary appeal that communism held for many in the 1930s.

ECONOMIC WARFARE AND ECONOMIC GLOBALIZATION

Perhaps the most significant difference between the two conflicts lies in the role of economic warfare and the constraints imposed by economic globalization.

The Spanish Civil War occurred in an era of relatively limited economic interdependence. Trade relationships, while important, could be severed without catastrophic economic consequences for the major powers. Financial systems were largely national, and energy dependencies were minimal compared to today's interconnected global economy.

The Ukraine conflict has been shaped fundamentally by

economic warfare on a scale unprecedented in modern history. Western sanctions against Russia have targeted not just military and political figures but entire sectors of the Russian economy, including exclusion from the SWIFT banking system and freezes on central bank assets. Russia's weaponization of energy exports, particularly natural gas to Europe, has created economic vulnerabilities that didn't exist in the 1930s.

These economic dimensions have complicated coalition formation. Countries like India and Turkey have maintained relationships with both sides, balancing geopolitical alignments with economic necessities. China's position reflects similar calculations whereby they provide diplomatic cover for Russia while trying to avoid sanctions that would damage Chinese access to Western markets and technology.

MILITARY INNOVATION AND TECHNOLOGICAL TESTING

Both conflicts served as laboratories for military innovation, though the pace and nature of technological development has accelerated dramatically.

Spain saw the testing of new aircraft designs, tank tactics, and coordination between different military branches. The lessons learned influenced military doctrine development that shaped World War II. However, the technological gaps between different systems were manageable, and innovations could be countered with existing technologies and tactics.

Ukraine has become a showcase for emerging technologies that are reshaping warfare. Drone swarms, artificial intelligence-assisted targeting, cyber warfare capabilities, and electronic warfare systems are being tested and refined in real combat conditions. The conflict has demonstrated the vulnerability of traditional military assets to relatively inexpensive precision munitions, potentially revolutionizing military planning worldwide.

The speed of innovation has also increased dramatically. New drone designs, software updates, and tactical adaptations can be deployed within weeks rather than years. This has created a continuous cycle of innovation and counter-innovation that keeps both sides constantly adapting their approaches.

INTERNATIONAL INSTITUTIONAL CONTEXT

The institutional frameworks governing international relations have evolved dramatically between the two conflicts, fundamentally altering how international coalitions form and operate.

The Spanish Civil War occurred in the dying days of the League of Nations system, with international institutions proving largely ineffective at managing great power competition. The Non-Intervention Committee became a symbol of institutional failure rather than effective diplomatic mediation.

The Ukraine conflict unfolds within a complex web of international institutions, NATO, the EU, the UN, the G7, and numerous other multilateral organizations. These institutions have provided frameworks for coordinating coalition responses, though they have also constrained the speed and scope of some responses due to consensus requirements and bureaucratic processes.

NATO's role has been particularly significant, providing a ready-made alliance structure for coordinating military assistance while maintaining the legal fiction that NATO itself is not a party to the conflict. The EU's coordination of sanctions demonstrates how modern institutional frameworks can enable rapid, comprehensive economic warfare in ways that were impossible in the 1930s.

REGIONAL AND GLOBAL POWER DYNAMICS

The broader international power structure has shifted fundamentally between these two conflicts, affecting coalition dynamics and strategic calculations.

In the 1930s, the international system was still multipolar, with several European powers maintaining significant global influence alongside the emerging superpowers of the United States and Soviet Union. This allowed for more fluid alliance patterns and gave smaller powers greater agency in choosing sides.

The Ukraine conflict occurs in a transitional moment in international relations when the unipolar moment of American hegemony is fading, but a stable multipolar system has not yet emerged. China's rise, Russia's revisionism, and the relative decline of European power have created new dynamics that don't map neatly onto historical precedents.

This transition has complicated coalition formation. Traditional allies sometimes have divergent interests, Turkey's position in NATO while maintaining relationships with Russia exemplifies these tensions. Emerging powers like India and Brazil have charted independent courses that reflect their own regional priorities rather than alignment with either major coalition.

LESSONS AND IMPLICATIONS

The comparison between these conflicts reveals both the persistence of certain patterns in international relations and the dramatic evolution of the mechanisms through which major power competition unfolds.

The tendency for local conflicts to become global proxy wars appears constant where geography and ideology combine to draw in external powers seeking to advance their interests and test their capabilities. However, the methods of involvement have become more sophisticated and less direct, reflecting both technological capabilities and the constraints of nuclear deterrence.

The role of economic interdependence as both a constraint on conflict and a weapon within it represents perhaps the most significant evolution. Modern conflicts must account for complex economic relationships in ways that were unnecessary in earlier eras.

The speed of technological innovation and adaptation has accelerated dramatically, creating more dynamic conflicts where advantages can shift rapidly. This has implications for military planning, alliance structures, and the duration of conflicts.

Finally, the institutional frameworks governing international relations, while more developed than in the 1930s, still struggle to manage major power competition effectively. The UN Security Council's paralysis during the Ukraine conflict echoes the League of Nations' ineffectiveness during the Spanish Civil War, suggesting that institutional evolution has not kept pace with the changing nature of international conflict.

CONCLUSION

The coalitions that formed around the Spanish Civil War and the Ukraine conflict reveal both the enduring patterns of major power competition and the dramatic evolution of international relations over the past century. While the fundamental dynamics of alliance formation, ideological competition, and proxy warfare remain recognizable, the mechanisms of involvement, the role of economic factors, and the pace of technological change have transformed how such conflicts unfold.

Understanding these parallels and differences is crucial for policymakers navigating current challenges and anticipating future conflicts. The Spanish Civil War served as a preview of World War II; the Ukraine conflict may similarly be shaping the contours of future major power competition in ways we are only beginning to understand. The coalitions forming today around this conflict will likely influence international relations for the period ahead, just as the alliances and enmities forged in Spain echoed through the global conflicts that followed.

The study of these historical parallels reminds us that while the tools and methods of international conflict evolve, the fundamental human dynamics of power, ideology, and alliance remain constants in the international system.

How contemporary leaders navigate these dynamics will determine whether the current moment leads toward greater international cooperation or slides toward the kind of global conflict that the Spanish Civil War foreshadowed in the 1930s.

ABOUT THE AUTHOR

Dr. Robbin Laird is a defense analyst and strategic researcher with more than four decades of experience examining military transformation and international security.

As Editor and Co-Founder of *Second Line of Defense* and *Defense.info*, and member of the Board of Contributors for Breaking Defense, he has established himself as an independent voice challenging conventional defense thinking.

Dr. Laird's engagement with the region predates the current crisis by three decades. In the early 1990s, he conducted the first Pentagon study on the newly independent Slavic states and traveled extensively throughout Ukraine, Russia, and Belarus during the pivotal post-Soviet transition. This early groundwork provided him unique insight into the forces that would eventually collide in 2022.

His career spans teaching positions at Columbia University (where he studied under Zbigniew Brzezinski), Princeton, and Johns Hopkins, alongside government work at the Defense Intelligence Agency and then operational work with the Center for Naval Analyses and Institute for Defense Analyses. This combination of academic rigor and operational understanding informs his analysis of how Ukraine's defense transformation is reshaping global military innovation and alliance structures.

Dr. Laird has spent recent years conducting field research across Europe, Australia, and North America, interviewing military leaders, defense ministers, and operational commanders as they grapple with the strategic earthquake triggered by Russia's invasion.

His work examines not just the war itself, but its cascading effects on defense industrial base transformation, alliance reconfiguration, and the emergence of new security architectures from the Indo-Pacific to the Arctic.

Operating under General Patton's principle that "if everyone is thinking alike, someone isn't thinking," Dr. Laird brings an essential contrarian perspective to defense analysis, one grounded in technological reality and strategic history rather than theoretical abstraction.

PUBLISHED BOOKS ON THE SOVIET UNION

AUTHORED BOOKS

- Hoffmann, Erik P., and Robbin F. Laird. 1982. "The Scientific-Technological Revolution" and Soviet Foreign Policy. New York: Pergamon Press. One of the early systematic studies linking Soviet conceptions of a scientific-technological revolution to changes in doctrine, strategy, and external behavior.
- Hoffmann, Erik P., and Robbin F. Laird. 1982. The Politics of Economic Modernization in the Soviet Union. Ithaca, NY: Cornell University Press. Analyzes the domestic political battles surrounding Soviet efforts to modernize the economy and the implications for policy effectiveness and reform.
- Laird, Robbin F., and Dale R. Herspring. 1984. The Soviet Union and Strategic Arms. Boulder, CO: Westview Press. Examines Soviet strategic nuclear doctrine, force development, and arms control policy to explain how Moscow approached the superpower nuclear balance.

- Hoffmann, Erik P., and Robbin F. Laird. 1985. Technocratic Socialism: The Soviet Union in the Advanced Industrial Era. Durham, NC: Duke University Press. Explores the emergence of a "technocratic socialism" in the USSR, arguing that advanced industrial and scientific imperatives were reshaping Soviet governance and priorities.
- Laird, Robbin F. 1986. The Soviet Union, the West and the Nuclear Arms Race. London: Croom Helm. Assesses Soviet nuclear strategy and arms competition with the West, emphasizing how Moscow viewed deterrence, parity, and the political uses of nuclear power.
- Laird, Robbin F., and Susan L. Clark. 1990. The USSR and the Western Alliance. Boston: Unwin Hyman. Evaluates Soviet strategy toward NATO and the evolving Western alliance, with particular attention to transatlantic cohesion and alliance vulnerabilities in the late Cold War.
- Laird, Robbin F. 1991. The Soviets, Germany, and the New Europe. Boulder, CO: Westview Press. Analyzes Soviet policy toward a changing Germany and European order at the end of the Cold War, focusing on how Moscow sought to adapt to unification and systemic transformation.
- Laird, Robbin F. 2019. France, the Soviet Union, and the Nuclear Weapons Issue. London: Routledge. Reconstructs the nuclear dimension of Franco–Soviet relations, showing how French and Soviet strategic choices intersected and diverged across the Cold War.

EDITED AND CO-EDITED SOVIET FOREIGN POLICY VOLUMES

- Laird, Robbin F., ed. 1986. Soviet Foreign Policy in a Changing World. New York: Free Press. A collected volume that traces the evolution of Soviet foreign policy across regions and themes, highlighting doctrinal shifts, regional strategies, and the impact of leadership change.

- Laird, Robbin F., ed. 1987. Soviet Foreign Policy. Proceedings of the Academy of Political Science, Vol. 36, No. 4. New York: Academy of Political Science. A proceedings issue bringing together analyses of Gorbachev-era "new thinking" and its domestic and international drivers at a pivotal moment in Soviet diplomacy.
- Fleron, Frederic J., Erik P. Hoffmann, and Robbin F. Laird, eds. 1991. Soviet Foreign Policy: Classic and Contemporary Issues. New York: Aldine de Gruyter. A comprehensive reader that juxtaposes foundational texts and contemporary analyses to illuminate continuity and change in Soviet foreign policy from 1917 to the Gorbachev period.

UNDERSTANDING THE GLOBAL CHANGE IN OUR TIME SERIES: A READER'S GUIDE

In an era when the ground beneath the international order seems to shift with each news cycle, the *Global Change in Our Time* series offers something increasingly rare: coherent analysis forged not in academic abstractions but through sustained real-world observation of how power actually operates in the twenty-first century. This ambitious collection of books addresses a deceptively simple question that confounds most contemporary analysts: how did we arrive at this moment of profound global transformation, and what does it mean for liberal democracies facing an ascendant authoritarian axis?

The series emerges from fifteen years of strategic analysis published on *Second Line of Defense* and *Defense.info*, platforms that have distinguished themselves by privileging operational reality over Washington consensus. Rather than recycling the comfortable assumptions that dominate establishment thinking, these works document how the world has fundamentally changed while American foreign policy elites remained trapped in what the series calls "mental amber" or frozen in the assumptions of the unipolar moment that followed the Soviet collapse and the tactical mindset of post-9/11 counterterrorism.

THE CORE THESIS: FROM UNIPOLARITY TO AUTHORITARIAN MULTIPOLARITY

At the heart of the series lies a diagnostic of historic importance: the liberal democratic "rules-based order" is not merely challenged but is being systematically replaced by an alternative architecture built by authoritarian powers who learned to operate outside Western frameworks while Western nations were distracted by asymmetric conflicts in the Middle East and convinced themselves that globalization would inevitably produce democratic convergence.

This is not the multipolar world of Cold War imagination, where competing ideological blocs faced off in clear opposition. Instead, the series documents the emergence of what it terms the "multi-polar authoritarian world", a sophisticated network of states that have constructed parallel financial systems, trade networks, and political alliances specifically designed to make Western sanctions ineffective and Western diplomatic pressure irrelevant. China and Russia didn't simply reject the American-led order; they built a shadow empire complete with alternative institutions that offer developing nations viable pathways to prosperity and security without requiring democratic reforms or adherence to Western human rights standards.

The uncomfortable reality documented across these volumes is that while America fought the war on terror and celebrated the spread of market economics as synonymous with liberal democracy, authoritarian capitalist powers created something far more durable and dangerous than Soviet communism ever achieved. They proved that economic development, technological advancement, and global influence require neither political freedom nor adherence to the norms that Western nations assumed were universal prerequisites for modern success.

DOCUMENTING ADMINISTRATIVE RESPONSES: FROM OBAMA TO BIDEN

The series provides systematic examination of how consecutive American administrations grappled with this transformation, revealing a persistent pattern of strategic confusion masked by confident rhetoric. The Obama administration volumes trace year-by-year how an administration that received the Nobel Peace Prize for "extraordinary efforts to strengthen international diplomacy" confronted the gap between its aspirations and the unforgiving realities of Russian resurgence and Chinese ambition. As Professor Kenneth Maxwell notes in his foreword, the Nobel committee may have "confused words with future performance", a confusion that extended far beyond Oslo to pervade American strategic thinking throughout the period.

The Biden administration analysis, titled *The Biden Administration Confronts Global Change: Déjà vu All Over Again*, examines whether "America is back" was genuine restoration or simply another slogan colliding with reality. Drawing from four years of real-time analysis spanning 2021-2024, this volume dissects how Biden's team discovered that the world had fundamentally changed during America's inward turn. The promise to "repair our alliances and engage with the world once again" confronted harsh truths: China wouldn't be contained by diplomatic niceties, Russia proved willing to wage total war in Europe, allies had learned to doubt American staying power, and the global order was fracturing faster than multilateral institutions could adapt.

These administrative assessments are not partisan exercises in blame assignment but rather clinical examinations of how American leadership regardless of party has struggled to comprehend and respond to systemic transformation. The series suggests that both Obama's sophisticated multilateralism and Biden's alliance restoration project operated from premises that no longer matched the world as it actually existed. Traditional American diplomacy, forged in an era of uncontested dominance, proved inadequate for navigating a reality where American power faces genuine rivals operating from fundamentally different assumptions about how international relations should function.

THE INTELLECTUAL ARCHITECTURE: STRATEGIC THINKERS CONFRONTING CHANGE

What distinguishes this series from conventional policy analysis is its method: rather than imposing theoretical frameworks from Washington think tanks, these volumes amplify the voices of strategic practitioners and unconventional thinkers who see patterns invisible to establishment consensus.

Assessing Global Change: Strategic Perspectives of Dr. Harald Malmgren presents the synthetic thinking of a figure the series describes as "a strategic chess grandmaster operating on the global board where nations, corporations, and ideologies battle for advantage." Malmgren doesn't just analyze the intersection of economics, diplomacy, and military power, he helped architect the modern global trading system that shapes how power flows between nations. His counsel reached presidents and corporate boardrooms, bridging the gap between economic theory and geopolitical reality. Former Air Force Secretary Michael Wynne observes that Malmgren's work provides "the architecture of the modern integration of economics, diplomacy, and statecraft", not academic theory but the blueprint for how advanced nations actually compete and survive.

The Malmgren volume, spanning fifteen years of assessments, reveals how a master strategist thinks several moves ahead while others remain trapped reacting to the last crisis. In a world where traditional boundaries between economic competition, diplomatic maneuvering, and national security have dissolved completely, his integrated approach offers the strategic clarity that leaders desperately need but rarely receive from conventional sources.

America, Global Military Competition, and Opportunities Lost: Reflections on the Work of Michael W. Wynne chronicles the visionary leadership of the 21st Secretary of the Air Force, whose prescient warnings and bold innovations were dismissed only to become America's most urgent military priorities two decades later. At the heart of Wynne's philosophy lies a doctrine that should haunt contemporary defense planners: "If you are ever involved in a fair fight, it is the result of poor planning."

This principle drove his relentless pursuit of military superiority through breakthrough technologies that would define 21st-century warfare: fifth-generation air dominance, hypersonic weapons, revolutionary logistics systems, cyber warfare capabilities, and the military "kill web" that transforms decision-making speed. His vision extended beyond platforms to forge interoperable coalitions and partnerships that multiply American power globally.

The 2008 firing of both Wynne and Air Force Chief of Staff Moseley represents, in the series' analysis, a catastrophic strategic blunder that symbolized America's retreat from the very capabilities essential for great power competition. Instead of maintaining air and naval superiority, the nation squandered two decades and countless resources on land-based operations that left America vulnerable and unprepared precisely when rivals advanced. When President Trump declared the arrival of "great power competition," the innovations Wynne had pioneered suddenly became national priorities but twenty years had been lost, years America couldn't afford to waste.

The Wynne volume stands as both tribute to transformational leadership and stark reminder of the price of strategic shortsightedness. It documents how institutional resistance to innovation, bureaucratic inertia, and the seductive pull of conventional wisdom can derail precisely the thinking that nations need most during periods of historical transformation.

HISTORICAL PERSPECTIVE: THE PHILOSOPHE EXAMINING MODERNITY

Kenneth Maxwell on Global Trends: An Historian of the 18th Century Looks at the Contemporary World provides the series with crucial historical depth. Maxwell, a distinguished historian of Brazil and the Iberian Peninsula, applies the lens of 18th-century philosophical inquiry to contemporary events, offering what the series describes as "a comprehensive perspective on the modern world as seen through the eyes of a professional historian."

Maxwell represents the tradition of the philosophes, scholars with deep expertise in specific subjects but maintaining wide-ranging perspectives on human affairs. His essays on the changing global order, viewed through the framework of someone who has studied previous periods of systemic transformation, provide unusual insights into what patterns recur across historical ruptures and what elements of our current moment represent genuinely novel challenges.

This historical dimension prevents the series from falling into presentism, the assumption that current crises are unprecedented and that past experience offers no guidance. Maxwell's work suggests that understanding how previous orders collapsed and were replaced can illuminate the dynamics we're witnessing, even as the specific technologies and ideologies differ dramatically from earlier eras.

REGIONAL FOCUS: FRENCH DEFENSE POLICY AND EUROPEAN RESPONSES

French Defense Policy Under President Macron: 2017-2021 extends the series' analysis to how European allies have responded to the transformed security environment. France, as a nuclear power with independent strategic capabilities and a tradition of autonomous defense thinking, provides a crucial test case for how liberal democracies are adapting to the reality of Russian revanchism and Chinese global ambition.

The volume examines how Macron's government reset French defense policy in response to these challenges, moving beyond the assumptions that guided European security during the post-Cold War peace dividend era. As Professor Maxwell notes in the foreword, the book provides "a comprehensive examination of the trials and tribulations and the success and failures and the illusions and delusions" of Macron's approach, an assessment that applies broadly to European efforts to develop strategic autonomy while maintaining transatlantic partnerships.

This European perspective reveals how the transformation documented in the series affects not just American policy but the entire constellation of liberal democracies that assumed the Cold War's end had permanently resolved fundamental questions about power, sovereignty, and security in the developed world.

THE CENTRAL CHALLENGE: PRESERVING LIBERAL DEMOCRACY IN AN AUTHORITARIAN AGE

Across all these volumes runs a single existential question: how can liberal democracies preserve the "rules-based order" and defend Western values when authoritarian powers have successfully constructed viable alternatives that operate by fundamentally different rules?

The series documents with unsparing clarity how ill-prepared democratic nations have been for this challenge. Western sanctions that once carried devastating economic consequences now "barely dent authoritarian economies." Russia sells oil and gas despite "extreme measures" to stop it. China, Iran, and North Korea operate their own financial networks largely immune to Western pressure. The Global South increasingly sees authoritarianism as a viable alternative to democracy, undermining the assumption that political freedom and economic development necessarily coincide.

The uncomfortable questions the series poses refuse easy answers:

- How did Western intelligence and policy communities miss the construction of this parallel world order while it was being built?

- Why do Western economic weapons no longer have the effectiveness they possessed even two decades ago?
- What happens when half the world stops playing by Western rules and proves that alternative systems can deliver prosperity and security?
- Can democracies adapt to a multipolar reality they didn't see coming and weren't prepared to confront?

These are not rhetorical questions posed to score ideological points. They represent genuine strategic dilemmas that liberal democracies must solve if they intend to survive as the dominant model for organizing advanced societies.

METHODOLOGY: REAL-TIME ANALYSIS AND PRACTITIONER ENGAGEMENT

What gives the series its distinctive character and analytical power is its methodology. These are not works produced in isolation by academics constructing theoretical models; they emerge from sustained engagement with military practitioners, defense officials, strategic thinkers, and alliance partners across continents. The analysis draws on interviews with senior officers, visits to operational units, participation in defense exercises, and continuous dialogue with allied defense establishments in Europe, Asia, and across the Pacific.

This practitioner focus means the series captures how transformation looks from the perspective of those actually implementing policy and operating military capabilities, rather than how it appears from Washington conference rooms. The analytical framework privileges operational reality over bureaucratic consensus, direct observation over official talking points, and the perspectives of those who must make strategies work in practice over the theories of those who merely advocate for them.

The series contributors include not just analysts but active-duty and retired flag officers, defense officials, academic specialists, and strategic thinkers from multiple nations. This diversity of perspective prevents the groupthink that characterizes so much defense analysis, where narrow circles recirculate the same assumptions until they acquire the status of unquestionable truth.

STRATEGIC IMPLICATIONS: FROM CRISIS MANAGEMENT TO CHAOS MANAGEMENT

A recurring theme across the series is the inadequacy of "crisis management" frameworks for the contemporary environment. Traditional crisis management assumes disruptions are temporary aberrations from normal stability that skilled diplomacy and appropriate force application can restore equilibrium and return systems to their proper functioning.

The series argues that we have transitioned from an era of crisis management to one requiring "chaos management" or the capacity to operate effectively within permanent uncertainty rather than seeking to restore stability that no longer exists. This represents a fundamental shift in how military and civilian organizations must function. Rather than building systems optimized for efficiency during peacetime with surge capacity for temporary crises, organizations must develop adaptive capacity for operating continuously within complexity.

This concept appears throughout the series' examination of military transformation, particularly in analyses of how services like the U.S. Marine Corps evolved under Force Design 2030, pivoting from counterinsurgency operations to strategic competition. New platforms like the MV-22 Osprey, CH-53K King Stallion, and F-35 Lightning II represent not merely technological upgrades but fundamental shifts toward distributed, network-centric operations where traditional hierarchies give way to more fluid command structures.

The shift from "kill chains" to "kill webs" exemplifies this transformation: rather than sequential processes where each step must complete before the next begins, modern operations require networked architectures where multiple actors can simultaneously detect, decide, and engage threats. This demands fundamentally different approaches to command and control, training, and operational concepts.

THE PRICE OF COMPLACENCY: LOST DECADES AND STRATEGIC SHORTSIGHTEDNESS

Perhaps the series' most damning finding is its documentation of how much time Western nations wasted while authoritarian powers prepared for strategic competition. The dismissal of Wynne's innovations, the continuation of land-centric operations long after they ceased serving strategic interests, the failure to invest in capabilities essential for maritime and air superiority, the persistent belief that economic interdependence would constrain authoritarian behavior, all represent what the series characterizes as catastrophic strategic shortsightedness.

Twenty years were lost that cannot be recovered. During those two decades, China built a navy that rivals American strength in its home waters, developed anti-access/area denial capabilities that make intervention in East Asia enormously costly, and established economic and diplomatic networks across Africa, Latin America, and Asia that provide alternatives to Western institutions. Russia reconstituted its military, developed hybrid warfare capabilities that exploit democratic vulnerabilities, and proved willing to use force to revise borders in Europe. Iran and North Korea advanced their missile and nuclear programs despite sanctions and diplomatic pressure.

Meanwhile, Western militaries optimized themselves for counterinsurgency operations against non-state actors, allowed readiness to decline, failed to modernize key capabilities, and assumed that technological superiority would remain constant without sustained investment. The comfortable belief that the "end of history" had arrived that liberal democracy represented the final form of human government and that authoritarianism was a dying anachronism prevented serious preparation for the competition that was already underway.

LOOKING FORWARD: STRATEGIC CLARITY IN AN UNFORGIVING WORLD

The series offers no illusions that these challenges can be easily overcome or that Western nations can simply return to the dominance they enjoyed during the unipolar moment. The world that existed from 1991 to approximately 2008 represented a historical anomaly, a brief period when American power faced no genuine rivals and liberal democracy seemed ascendant. That world is gone, and nostalgia for it serves no strategic purpose.

Instead, the series calls for hard-headed assessment of how liberal democracies can compete effectively in a multipolar environment where authoritarian powers have proven they can deliver economic growth, technological advancement, and national security without political freedom. This requires abandoning comfortable assumptions about the inevitable superiority of democratic systems and instead demonstrating through performance that free societies can outcompete authoritarian alternatives.

This demands several fundamental shifts: from seeking stability to building adaptive capacity for operating in continuous complexity; from assuming Western institutions will remain globally dominant to recognizing that parallel systems now offer viable alternatives; from believing economic interdependence constrains conflict to understanding that authoritarian states have insulated themselves from Western economic pressure; from expecting allies to automatically align with American preferences to recognizing that alliance management requires continuous effort and credible commitment.

Most fundamentally, it requires what the series' founding principle demands: independent thinking that challenges consensus. As the series notes in invoking General Patton's observation, "if everyone is thinking alike, someone isn't thinking." The establishment consensus that guided Western policy for the past two decades has failed spectacularly to anticipate or effectively respond to the transformation it documents. Continuing to rely on that consensus promises continued failure.

CONCLUSION: ESSENTIAL READING FOR UNDERSTANDING OUR HISTORICAL MOMENT

The Global Change in Our Time series represents an indispensable resource for anyone seeking to understand the systemic transformation reshaping global order. Rather than offering comfortable narratives about Western resilience or inevitable democratic triumph, these volumes provide unflinching analysis of how liberal democracies arrived at this moment of genuine peril and what it will take to navigate successfully through the competition that defines our era.

The series bridges the gap between academic theory and operational reality, between policy prescription and practical implementation, between American perspectives and allied viewpoints. It amplifies voices that mainstream discourse often ignores, privileges direct observation over received wisdom, and refuses to mistake confident rhetoric for strategic competence.

For policymakers, military professionals, strategic analysts, and informed citizens trying to make sense of a world that seems increasingly chaotic and threatening, these volumes offer the strategic clarity that conventional analysis so often fails to provide. They document not just what has changed but how it changed, who saw it coming, and what price has been paid for ignoring those warnings.

Most importantly, the series makes clear that the challenges facing liberal democracies are not temporary disruptions but system-defining transformations that will determine whether democratic governance can survive in competition with authoritarian alternatives that have proven more capable and resilient than anyone in the West predicted. The answer to that existential question remains to be written, but understanding how we arrived at this moment which these volumes provide with exceptional clarity represents the essential first step toward writing that answer successfully.

In an era when strategic shortsightedness has become the norm and comfortable assumptions substitute for rigorous analysis, the *Global Change in Our Time* series stands as a model of what serious strategic thinking requires: willingness to challenge consensus, attention to operational reality over bureaucratic preference, integration of multiple perspectives and disciplines, and above all, the intellectual courage to acknowledge uncomfortable truths about the world as it actually exists rather than as we wish it to be.

NOTES

Prologue

1. https://apps.dtic.mil/sti/citations/ADA268856
2. https://defenceviewpoints.co.uk/articles-and-analysis/russia-iran-and-the-biden-speech
3. https://www.politico.com/blogs/politico-now/2014/03/kerry-russia-behaving-like-its-the-19th-century-184280

Introduction

1. https://www.rferl.org/a/ukraine-north-korea-retreat-casualties-russia/33298570.html
2. https://www.npr.org/2025/04/28/nx-s1-5379436/north-korea-russia-ukraine-troops-putin-kim
3. https://www.nbcnews.com/world/russia/north-korea-thousands-troops-russia-terrorize-ukraine-war-rcna210111
4. https://www.newsweek.com/putin-kim-xi-russia-north-korea-china-strategic-alliance-2066246
5. https://www.gov.uk/government/news/statement-of-the-coalition-of-the-willing-meeting-by-the-leaders-of-the-united-kingdom-france-and-ukraine-10-july-2025
6. https://thedefensepost.com/2024/12/24/nato-organization-supporting-ukraine/
7. https://www.cer.eu/publications/archive/policy-brief/2025/can-europe-save-ukraine-itself-putin-trump; https://www.nytimes.com/2025/10/20/world/europe/ukraine-russia-war-europe.html
8. https://pism.pl/publications/japan-increases-its-support-for-ukraine; https://www.president.gov.ua/en/news/finansova-pidtrimka-ta-vidbudova-ukrayini-vidbulasya-zustric-98509
9. https://www.airuniversity.af.edu/JIPA/Display/Article/4266915/japan-the-war-in-ukraine-and-japaneurope-relations-a-g7nato-alignment-perspecti/
10. https://www.rusi.org/explore-our-research/publications/commentary/japans-defense-budget-surge-new-security-paradigm
11. https://en.yna.co.kr/view/AEN20240711010600315; https://www.reuters.com/world/asia-pacific/south-korea-nato-boost-partnership-security-cyber-threats-2023-07-11/
12. https://thediplomat.com/2024/11/south-koreas-deepening-dilemma-over-ukraine/; https://www.npr.org/2024/10/24/nx-s1-5163246/south-korea-weapons-ukraine-north-korea-troops-russia
13. https://www.csis.org/analysis/russias-battlefield-woes-ukraine
14. https://www.csis.org/analysis/russias-battlefield-woes-ukraine
15. https://www.themoscowtimes.com/2025/07/29/we-could-lose-it-in-60-days-russia-closes-in-on-ukrainian-stronghold-of-pokrovsk-a90004

16. https://www.cnn.com/2025/01/31/europe/ukraine-russia-kursk-north-korean-troops-intl/index.html
17. https://www.gov.uk/government/news/new-coalition-of-the-willing-headquarters-as-leaders-step-up-support-for-ukraines-immediate-flight

1. The Putin Dynamic: The Prelude to the War

1. Angus Roxburgh, *The Strongman: Vladimir Putin and the Struggle for Russia* (I.B. Tauris and Co., 2012).
2. Celestine Bohlen, "A.A. Sobchak Dead at 62; Mentor to Putin," The New York Times, February 21, 2000), https://www.nytimes.com/2000/02/21/world/aa-sobchak-dead-at-62-mentor-to-putin.html
3. https://unherd.com/2022/01/what-the-west-gets-wrong-about-putin/.
4. Steven Lee Myers, *The New Tsar: The Rise and Reign of Vladimir Putin* (New York: Knopf Doubleday Publishing Group. Kindle Edition, 2016), 143-144.
5. http://archive.premier.gov.ru/eng/premier/press/ru/799/print/
6. https://sldinfo.com/2019/09/the-putin-narrative-and-the-way-ahead-for-russia/
7. Vladimir Putin, "Speech and the Following Discussion at the Munich Conference on Security Policy," February 10, 2007. http://en.kremlin.ru/events/president/transcripts/24034
8. Brian D. Taylor, *The Code of Putinism* (Oxford University Press. Kindle Edition, 2018), 192.
9. Shaun Walker, *The Long Hangover* (Oxford University Press, 2018, Kindle Edition).
10. Shaun Walker, *The Long Hangover* (Oxford University Press, 2018, Kindle Edition), 246.
11. Dmitri Trenin, *What Is Russia Up to in the Middle East?* (New York: Wiley, Kindle Edition, 2018),
12. Brian Taylor, *The Code of Putinism*. (Oxford University Press, 2018).
13. Tony Wood, *Russia Without Putin: Money, Power and the Myths of the New Cold War* (Verso Books, 2018).

2. The Run-up to Putin's 2022 Invasion of Ukraine

1. https://www.staradvertiser.com/2021/06/22/breaking-news/russia-details-exercise-near-hawaii-to-destroy-a-carrier-strike-group/
2. https://www.twz.com/41197/russia-practices-destroying-enemy-carriers-in-pacific-drills-sending-u-s-alarm-bells-ringing
3. https://euromaidanpress.com/2021/07/10/information-warfare-at-sea/
4. https://jamestown.org/program/russia-bungles-pre-planned-intercept-of-uk-navy-vessel-off-coast-of-crimea/
5. https://defense.info/global-dynamics/2021/09/central-asia-on-the-front-lines/
6. https://www.wsj.com/articles/cause-ukraine-war-robert-service-moscow-putin-lenin-stalin-history-communism-invasion-kgb-fsb-11646413200?mod=Searchresults_pos1&page=1
7. https://bidenwhitehouse.archives.gov/briefing-room/statements-releases/2021/09/01/joint-statement-on-the-u-s-ukraine-strategic-partnership/
8. https://www.state.gov/u-s-ukraine-charter-on-strategic-partnership/

9. https://sldinfo.com/2021/11/poland-faces-the-belarus-russia-migrant-battering-ram/

10. https://www.reuters.com/article/world/poles-recall-nazi-seizure-of-radio-station-see-parallels-with-today-idUSKBN0GT1TY/

1. The Path Not Taken

1. Selected sources on Russia and the PfP program. https://time.com/5564207/russia-nato-relationship/; https://www.yalejournal.org/publications/nato-partnership-for-peace; https://nsarchive.gwu.edu/briefing-book/russia-programs/2018-03-16/nato-expansion-what-yeltsin-heard; https://en.wikipedia.org/wiki/Russia%E2%80%93NATO_relations. A comprehensive look at the relationship was provided by Martin A. Smith, *Russia and NATO Since 1991* (2006). This book has been reissued by Routledge in its Cold War series, where several of my books can be found as well.

2. Strategic Miscalculations

1. https://www.thebarentsobserver.com/news/brain-drain-hammering-russia-more-than-2500-scientists-have-already-left-this-is-a-disaster-experts-say/102122

2. https://www.science.org/content/article/scientists-russia-struggle-world-transformed-its-war-ukraine

3. https://epjdatascience.springeropen.com/articles/10.1140/epjds/s13688-023-00389-3

4. https://www.wgbh.org/news/local/2022-02-25/mit-abandons-russian-high-tech-campus-partnership-in-light-of-ukraine-invasion

5. https://www.businessinsider.com/russia-blocked-access-cern-shows-ukraine-war-damaging-its-science-2024-10

6. https://epjdatascience.springeropen.com/articles/10.1140/epjds/s13688-023-00389-3

7. https://www.chathamhouse.org/2025/07/russias-struggle-modernize-its-military-industry/identifying-weaknesses-russias-military

8. https://lims.ac.uk/perspectives/lets-support-russian-scientists-at-the-expense-of-russian-science/

9. https://www.thetimes.com/comment/columnists/article/putins-hostility-to-scientists-is-causing-russian-brain-drain-72kh9tk5w

10. https://voxukraine.org/en/sanctions-against-the-russian-science-current-results-so-far

11. https://defense.info/global-dynamics/2025/08/russian-geopolitical-challenges-the-economic-relationship-with-the-baltic-states/

12. https://defense.info/global-dynamics/2025/07/russian-geopolitical-challenges-kazakhstans-strategic-pivot/

13. https://www.pravda.com.ua/eng/articles/2024/01/10/7436569/

14. https://euromaidanpress.com/2025/06/17/kazakhstan-cuts-explosives-supply-to-russia-ukraine-war/

15. https://defense.info/global-dynamics/2025/07/russian-geopolitical-challenges-kazakhstans-strategic-pivot/

16. https://defense.info/global-dynamics/2025/07/russian-geopolitical-challenges-azerbaijans-hardline-turn-against-russia/

17. https://www.washingtonpost.com/world/2025/07/03/russia-azerbaijan-relations-arrests/

18. https://www.abc.net.au/news/2025-10-10/azerbaijani-plane-crash-russian-air-defenses-to-blame-putin-says/105876934

19. https://www.aljazeera.com/opinions/2025/7/10/the-south-caucasus-is-slipping-from-russias-grasp

20. https://defense.info/global-dynamics/2025/07/russian-geopolitical-challenges-azerbaijans-hardline-turn-against-russia/

21. https://miwi-institut.de/archives/3400; https://defense.info/global-dynamics/2025/08/russian-geopolitical-challenges-economic-relations-with-germany/

22. https://www.weforum.org/stories/2022/08/energy-crisis-germany-europe/

23. https://defense.info/global-dynamics/2025/08/russian-geopolitical-challenges-economic-relations-with-germany/

24. https://interfax.com/newsroom/top-stories/109168/

25. https://defense.info/global-dynamics/2025/07/russian-geopolitical-challenges-a-window-into-sino-russian-territorial-tensions/

26. https://eurasianet.org/one-island-two-countries-a-look-at-how-chinese-russian-relations-are-playing-out-in-the-far-east

27. https://defense.info/global-dynamics/2025/07/russian-geopolitical-challenges-a-window-into-sino-russian-territorial-tensions/

28. https://sldinfo.com/2025/07/putins-war-economy-how-ukraines-invasion-became-a-tool-for-domestic-control/

29. https://sldinfo.com/2025/07/putins-war-economy-how-ukraines-invasion-became-a-tool-for-domestic-control/

30. https://defense.info/re-thinking-strategy/2025/08/putins-shadow-boxing-with-nato-how-russia-created-the-very-threats-it-claims-to-combat/

1. Moving Beyond the Soviet Military Legacy

1. https://www.csis.org/analysis/how-ukraine-rebuilt-its-military-acquisition-system-around-commercial-technology

2. https://www.sipri.org/commentary/topical-backgrounder/2025/transformation-ukraines-arms-industry-amid-war-russia

3. https://carnegieendowment.org/research/2023/12/arsenal-of-democracy-integrating-ukraine-into-the-wests-defense-industrial-base?lang=en

4. https://www.atlanticcouncil.org/blogs/ukrainealert/oleksii-reznikov-ukraines-defense-doctrine-will-define-countrys-future/

5. https://www.sipri.org/commentary/topical-backgrounder/2025/transformation-ukraines-arms-industry-amid-war-russia

6. https://www.commercialuavnews.com/how-ukraine-is-driving-the-drone-revolution; https://www.wired.com/story/ukraine-drone-startups-russia/

7. https://kyivindependent.com/turkish-drone-manufacturer-baykar-begins-construction-of-factory-near-kyiv/; https://www.lvivherald.com/post/ukraine-s-defense-industry-renaissance-from-soviet-legacy-to-nato-standards

8. https://nationalinterest.org/blog/buzz/how-drones-became-central-to-ukraines-military-strategy-hk

9. https://united24media.com/latest-news/ukraines-drone-surge-goes-full-throttle-900-production-boom-signals-shift-to-total-war-mode-9680

10. https://www.csis.org/analysis/ukraines-future-vision-and-current-capabilities-waging-ai-enabled-autonomous-warfare; https://www.kyivpost.com/post/55897

11. https://ukrainesarmsmonitor.substack.com/p/fpv-drone-localization-in-ukraine; https://www.forbes.com/sites/davidhambling/2025/04/08/ukraine-is-making-fpv-drones-without-chinese-parts-and-at-lower-cost/; https://militarnyi.com/en/news/ukraine-produces-first-thousand-fully-domestic-fpv-drones/

12. https://www.axios.com/2025/06/02/ukraine-spider-web-surprise-attack

13. https://www.news9live.com/technology/tech-news/ukraines-operation-spider-web-how-20-year-old-ardupilot-software-destroyed-russian-bombers-2862861

14. https://www.reuters.com/business/aerospace-defense/ukraine-stages-major-attack-russian-aircraft-with-drones-security-official-says-2025-06-01/

15. https://euromaidanpress.com/2025/06/03/meet-first-contacts-osa-the-ukraine-fpv-drone-used-to-strike-russian-bombers-in-spiderweb-operation/

16. https://www.kyivpost.com/post/53784

17. https://www.iceye.com/newsroom/press-releases/iceye-signs-contract-to-provide-government-of-ukraine-with-access-to-its-sar-satellite-constellation

18. https://www.newsweek.com/ukraine-gur-military-intelligence-agency-russia-iceye-sar-satellite-imagery-1977397

19. https://breakingdefense.com/2023/02/spurred-by-ukraine-war-18-western-coun tries-plan-to-share-remote-sensing-data/

20. https://chronicle.lu/category/space/44382-luxembourg-contributes-eur16-5m-to-allied-persistent-surveillance-from-space

21. https://www.iceye.com/newsroom/press-releases/iceye-signs-contract-to-provide-government-of-ukraine-with-access-to-its-sar-satellite-constellation

22. https://sciendo.com/2/v2/download/article/10.2478/raft-2024-0020.pdf

23. https://united24media.com/war-in-ukraine/how-starlink-became-ukraines-life line-in-war-5774

24. https://foreignpolicy.com/2022/11/22/ukraine-internet-starlink-elon-musk-russia-war/; https://irregularwarfare.org/articles/when-a-ceo-plays-president-musk-star link-and-the-war-in-ukraine/

25. https://euromaidanpress.com/2025/03/23/the-musk-factor-would-ukraines-front line-really-collapse-without-starlink/

26. https://www.armyrecognition.com/archives/archives-land-defense/land-defense-2024/ukrainian-defense-industry-ukroboronprom-enters-into-global-top-50-as-fastest-growing-defense-company

27. Ukroboronprom official website, accessed September 2025.

28. https://www.npr.org/2024/10/13/nx-s1-5147284/ukraine-drones-russia-war

29. https://kyivindependent.com/russia-says-european-satellites-aiding-ukraine-are-legitimate-targets-for-signal-jamming/

2. The Digital Arsenal: The Israeli-Ukrainian Partnership

1. https://www.lvivherald.com/post/military-links-between-ukraine-and-israel

2. https://www.i24news.tv/en/news/innov-nation/artc-innov-vtr-3

3. https://www.armyrecognition.com/news/aerospace-news/2024/israel-draws-on-ukraines-expertise-to-counter-iranian-drone-threats

4. https://pwrteams.com/content-hub/blog/articles/ukrainian-it-industry-during-the-war.

5. https://alcor-bpo.com/what-you-should-know-before-hiring-software-developers-in-ukraine/

6. https://idapgroup.com/blog/software-development-outsourcing-companies-in-ukraine/

7. https://sifted.eu/articles/ukrainian-tech-2024-brnd.

8. https://odessa-journal.com/israel-is-in-contact-and-establishing-connections-with-ukrainian-drone-manufacturers-to-foster-cooperation-and-exchange-expertise

9. https://dronelife.com/2025/11/12/ukraines-drone-boom-how-wartime-innovation-is-reshaping-the-global-drone-industry/.

10. https://news.usni.org/2024/04/18/ukraines-experience-in-developing-lethal-drones-should-be-lesson-for-nato-says-panel

11. https://www.csis.org/analysis/ukraines-future-vision-and-current-capabilities-waging-ai-enabled-autonomous-warfare.

12. https://www.404media.co/ukraines-massive-drone-attack-was-powered-by-open-source-software/

13. https://www.csis.org/analysis/ukraines-future-vision-and-current-capabilities-waging-ai-enabled-autonomous-warfare.

14. https://kyivindependent.com/better-late-than-never-israeli-companies-finally-reach-out-to-ukraine-to-help-fight-iranian-drones/

15. https://kyivindependent.com/better-late-than-never-israeli-companies-finally-reach-out-to-ukraine-to-help-fight-iranian-drones/

16. https://www.shomrim.news/eng/israel-ukraine-fight-iranian-drones.

17. https://themedialine.org/by-region/drowning-in-drones-israel-turns-to-ukraine-for-expertise/.

3. War in the Modern Age

1. https://www.tandfonline.com/doi/full/10.1080/08850607.2025.2522222

2. https://arxiv.org/html/2501.01884v3

3. https://time.com/6158437/telegram-russia-ukraine-information-war/

4. https://indianexpress.com/article/explained/russia-ukraine-war-telegram-app-7847165/

5. https://www.businessinsider.com/ukraine-military-e-enemy-telegram-app-2022-4

6. https://www.bohrium.com/paper-details/ukrainian-intelligence-s-use-of-telegram-in-wartime/1149639883883544610-13440

7. https://www.kyivpost.com/post/17868

8. https://www.researchgate.net/publication/377109559

9. https://united24media.com/war-in-ukraine/how-a-pro-russian-telegram-channel-monetizes-the-slaughter-of-ukrainians-2777

10. https://www.uttryckmagazine.com/2025/02/27/telegram-war-fuels-itself/

11. https://www.atlanticcouncil.org/event/digital-occupation-inside-russias-telegram-battle-in-ukraine/

12. https://www.egyptindependent.com/wagner-chief-admits-to-founding-firm-sanctioned-by-us-over-alleged-election-interference/
13. https://online.ucpress.edu/cpcs/article-abstract/doi/10.1525/cpcs.2025.2465591/211850/Creating-Good-Young-PatriotsRussian-Youth-Leaders?redirectedFrom=full text
14. https://meduza.io/en/feature/2025/01/16/we-need-eyes-and-ears
15. https://www.rferl.org/a/russia-gru-nazi-sabotage-recruitment-telegram/33539661.html; https://www.occrp.org/en/investigation/make-a-molotov-cocktail-how-europeans-are-recruited-through-telegram-to-commit-sabotage-arson-and-murder

4. Ukraine, Democratic Deficits, and Strategic Necessity

1. Freedom House, *Freedom in the World 2024: Ukraine* (Washington, DC: Freedom House, 2024). Ukraine rated "Partly Free" with score of 61/100
2. Transparency International, *Corruption Perceptions Index 2023* (Berlin: Transparency International, 2024). Ukraine ranked 104 out of 180 countries.
3. Michael Kimmage, "Ukraine Is Not a Democracy: The Case for Supporting It Anyway," *Foreign Affairs*, March 15, 2023.
4. Article 64, Constitution of Ukraine (1996, as amended). Provides for suspension of certain rights during martial law.
5. Vladimir Putin, "On the Historical Unity of Russians and Ukrainians," Official Website of the President of Russia, July 12, 2021. See also: Timothy Snyder, *The Road to Unfreedom: Russia, Europe, America* (New York: Tim Duggan Books, 2018).
6. Steven Levitsky and Lucan Way, *Competitive Authoritarianism: Hybrid Regimes After the Cold War* (New York: Cambridge University Press, 2010), 5-13.
7. Daron Acemoglu and James Robinson, *Why Nations Fail: The Origins of Power, Prosperity, and Poverty* (New York: Crown, 2012). See also: Samuel Huntington, *The Third Wave: Democratization in the Late Twentieth Century* (Norman: University of Oklahoma Press, 1991).

5. Ukraine's Coalition Building

1. https://mod.gov.ua/en/news/a-snapshot-of-ramstein-in-figures-25-meetings-and-over-145-billion-in-aid-to-ukraine
2. https://www.nato.int/cps/en/natohq/topics_37750.htm
3. https://ies.lublin.pl/en/comments/the-ramstein-group-prerequisites-for-creation-and-role-in-ensuring-ukraines-military-capabilities-part-1/
4. https://www.swp-berlin.org/publikation/diplomacy-in-the-context-of-the-russian-invasion-of-ukraine
5. https://www.atlanticcouncil.org/blogs/ukrainealert/exploring-the-secrets-of-ukraines-successful-wartime-diplomacy/
6. https://mod.gov.ua/en/news/30th-ramstein-how-the-ukraine-defense-contact-group-meetings-strengthen-ukraine-s-defense-capabilities
7. https://mod.gov.ua/en/news/ukraine-to-receive-new-aid-packages-following-ramstein-meeting

8. https://newsukraine.rbc.ua/analytics/us-won-t-attend-in-person-but-will-join-online-1744344374.html
9. https://www.aljazeera.com/news/2022/12/28/what-is-zelenskyys-10-point-peace-plan
10. https://www.kyivpost.com/post/23162
11. https://prismua.org/en/english-ukraines-peace-formula-global-dimension-and-shared-principles/
12. https://dif.org.ua/en/article/how-can-ukraine-reach-out-to-the-global-south-omar-ashour
13. https://spectator.clingendael.org/en/publication/new-allies-ukraines-diplomatic-battle-global-south; https://www.eurasiantimes.com/3rd-august-the-west-drags-global-south-into/
14. https://crimea-platform.org/en/international-conference-crimea-global-under standing-ukraine-through-the-south/
15. https://ecfr.eu/article/deconstructing-russias-anti-colonial-posturing-in-the-global-south/
16. https://www.aljazeera.com/news/2025/8/21/tracking-us-and-nato-support-for-ukraine-a-full-breakdown
17. https://ppr.lse.ac.uk/articles/10.31389/lseppr.88
18. https://cospr.org/diplomatic-maneuvers-and-balancing-power-implications-us-pursuit-of-peace-in-ukraine/

6. From Cinematic Battles to Real War

1. https://sldinfo.com/2025/08/from-cinematic-battles-to-real-war-ukraines-journey-from-waterloo-to-today/

1. Europe's Defense Renaissance

1. https://defense.info/re-thinking-strategy/2021/07/strategic-imagination-meeting-the-challenge/
2. https://defense.info/interview-of-the-week/the-perspective-of-brigadier-general-retired-rainer-meyer-zum-felde-on-the-future-of-german-defense/
3. https://defense.info/multi-domain-dynamics/2025/05/from-plywood-to-precision-how-german-drones-are-changing-ukraines-battlefield-strategy/
4. https://defense.info/re-shaping-defense-security/2025/07/from-ukraine-to-nato-how-germanys-drone-partnership-could-transform-alliance-defense/; https://sldinfo.com/2025/08/germanys-drone-revolution-partnership-with-ukraine/
5. https://www.dsei.co.uk/news/dsei-uk-2025-uk-mass-produce-ukrainian-intercep tor-drone
6. https://euromaidanpress.com/2025/06/03/ukraine-races-to-build-drone-intercep tors-as-russia-ramps-up-shahed-attacks/
7. https://www.president.gov.ua/en/news/ukrayina-vzhe-vikoristovuye-perehoplyu vachi-dlya-zbittya-sha-98561
8. https://english.nv.ua/nation/ukrainian-drones-go-global-with-major-uk-factory-and-test-site-50543100.html

9. https://www.euronews.com/my-europe/2025/05/28/germany-to-jointly-produce-long-range-weapons-with-ukraine-friedrich-merz-says; https://sldinfo.com/2025/08/germanys-drone-revolution-partnership-with-ukraine/
10. I discuss this in detail in my 2025 published book entitled: *Training for the High-End Fight: The Paradigm Shift for Combat Pilot Training*.
11. https://sldinfo.com/2025/08/germanys-drone-wall-how-ukraines-battlefield-innovation-impacts-on-natos-eastern-defense/
12. https://www.nato.int/cps/en/natohq/news_219119.htm
13. https://www.pravda.com.ua/eng/news/2024/05/25/7457580/
14. https://defense.info/re-shaping-defense-security/2025/07/germanys-defense-initiatives-building-europes-new-security-architecture/
15. https://sldinfo.com/2025/07/anglo-french-alliance-reborn-modernizing-defense-in-a-new-european-era/
16. https://www.osw.waw.pl/en/publikacje/analyses/2025-07-11/france-and-united-kingdom-beginning-bilateral-coordination-nuclear
17. https://www.lemonde.fr/international/article/2025/07/11/la-france-et-le-royaume-uni-resserrent-la-coordination-de-leurs-forces-nucleaires-face-a-la-menace-russe_6620577_3210.html
18. https://www.gov.uk/government/news/treaty-between-the-united-kingdom-of-great-britain-and-northern-ireland-and-the-federal-republic-of-germany-on-friendship-and-bilateral-cooperation
19. https://economictimes.indiatimes.com/news/defense/missiles-mutual-aid-and-a-drone-factory-uk-and-germany-sign-historic-defense-treaty-with-eye-on-russia-and-trumps-nato-drift/articleshow/122765303.cms?from=mdr
20. https://www.reuters.com/world/frances-macron-announces-plan-accelerate-military-spending-2025-07-13/. But then there is the reality: https://www.reuters.com/world/europe/french-fiscal-woes-crimp-macrons-defense-spending-plans-2025-03-05/
21. https://www.politico.eu/article/macron-to-eu-colleagues-stop-buying-american-buy-european/

3. The Northern Shield

1. https://www.wsj.com/world/europe/paramilitary-expansion-shows-scale-of-war-preparations-on-natos-eastern-frontier-11b145c3
2. https://www.wsj.com/world/europe/paramilitary-expansion-shows-scale-of-war-preparations-on-natos-eastern-frontier-11b145c3?gaa_at=eafs&gaa_n=AWEtsqdX poVMnDHSRVOTYsiLVeLJMwwCnWovyu5PYxRysHlEYN7pzRRW75HHHpE J3ck%3D&gaa_ts=692a30d9&gaa_sig= HR_Pj5gs_jY4owsT3SrVlv32YqYpzrbL4yMF7W6kipxGPrWJZIxysQYECn6W32X tQp-HdczR_cZtELhJZFEfAA%3D%3D
3. https://easternflank.org/estonian-defense-league-on-modern-territorial-defense-in-estonia/
4. https://e-estonia.com/rethinking-security-in-an-interconnected-era/
5. https://cepa.org/article/latvia-poised-to-return-to-conscription-as-russian-menace-grows/
6. https://www.defenseadvancement.com/resources/finnish-defence-forces/

7. https://united24media.com/latest-news/finland-confirms-50000-bomb-shelters-ready-to-protect-nearly-entire-population-12526

8. https://www.hertie-school.org/en/news/live-on-campus/detail/content/comprehensive-security-the-finnish-model-for-21st-century-threats

9. https://www.kas.de/en/country-reports/detail/-/content/finland-s-response-to-hybrid-threats-in-the-baltic-sea.   https://www.hybridcoe.fi/wp-content/uploads/2024/05/20240527-Hybrid-CoE-Working-Paper-31-Building-resilience-to-hybrid-threats-WEB.pdf

10. https://www.nytimes.com/1992/05/08/world/finland-in-3-billion-deal-for-64-mcdonnell-douglas-combat-jets.html

11. https://breakingdefense.com/2022/05/norways-chief-of-defense-finland-sweden-in-nato-opens-up-a-lot-of-possibilities/

12. https://sldinfo.com/2018/05/a-norwegian-perspective-on-cross-border-training-north/

13. Quote from Robbin Laird and Murielle Delaporte, *The Return of Direct Defense in Europe: Meeting the Challenge of the XXIst Century Authoritarian Powers.*

14. https://defense.info/re-shaping-defense-security/2022/07/finland-and-the-next-phase-of-european-defense/

15. https://breakingdefense.com/2025/11/with-sweden-baltic-sea-now-a-lake-full-of-nato-submarines/

16. https://capstone.ndu.edu/Portals/83/20-2%20EUR2%20Trip%20Book%20Vol%2011%20red.pdf

17. https://www.government.se/press-releases/2025/03/biggest-support-package-to-ukraine-so-far-increases-swedish-support-to-sek-29.5-billion-for-2025

18. https://breakingdefense.com/2025/03/warning-russia-has-initaitive-in-ukraine-sweden-announces-record-1-6b-defense-aid-package/

19. https://www.nato.int/en/news-and-events/articles/news/2025/01/13/nato-deputy-secretary-general-in-sweden-allies-must-step-up-support-for-ukraine

20. https://www.reuters.com/world/europe/norway-should-significantly-raise-its-aid-ukraine-pm-says-2025-03-06/

21. https://re-russia.net/en/analytics/0310/

22. https://www.norway.no/en/ukraine/norway-ukraine/nansen-programme-for-ukraine/

23. https://www.regjeringen.no/en/whats-new/norway-provides-air-defense-worth-seven-billion-kroner-to-ukraine/id3116609/

24. https://cepa.org/article/kremlin-snarls-as-norway-surges-ukraine-aid/

25. https://breakingdefense.com/2025/07/denmark-eyes-additional-f-35-fighter-jets-patriot-aid-breakthrough-for-ukraine/

26. https://www.defensenews.com/global/europe/2025/01/14/denmark-flies-home-more-f-35s-for-patrols-to-buffer-upgrade-delays/

27. https://thedefensepost.com/2025/05/19/denmark-f-35-jets-lockheed/

28. https://www.flightglobal.com/fixed-wing/denmark-eyes-greenland-airport-upgrades-to-support-f-35-fighter-deployment/161359.article

29. https://breakingdefense.com/2025/01/denmark-strengthens-arctic-defense-with-2b-package-for-naval-vessels-drones/

30. https://www.aljazeera.com/news/2025/1/28/denmark-to-pump-2bn-into-arctic-security-as-trump-eyes-greenland

31. https://www.youtube.com/watch?v=vb7uHQMchGs

32. https://www.newsweek.com/denmark-military-two-billion-boost-donald-trump-greenland-threat-2022685

33. https://defence-industry.eu/denmark-norway-and-sweden-pledge-500-million-in-support-for-ukraine-under-nato-initiative/

34. https://www.reuters.com/world/sweden-norway-denmark-give-500-million-nato-project-send-us-weapons-ukraine-2025-08-05/

35. https://valtioneuvosto.fi/en/-/236553176/minister-of-defence-hakkanen-nordic-and-baltic-countries-announce-usd-500-million-package-of-support-for-ukraine-under-purl

36. https://www.aa.com.tr/en/economy/military-spending-boost-to-give-poland-largest-army-in-eu/2801841;   https://cepa.org/article/poland-becomes-a-defense-colossus/

37. https://notesfrompoland.com/2024/08/12/poland-signs-1-2bn-deal-to-produce-launchers-for-patriot-missile-systems/

38. https://defence24.com/armed-forces/himars-deliveries-to-poland-finalised

39. https://nestcentre.org/escalation-in-europe/

40. https://www.lrt.lt/en/news-in-english/19/2742115/lrt-investigation-lithuanian-sim-cards-used-to-track-smuggling-balloons-from-belarus

41. https://www.amnesty.org/en/latest/news/2024/06/finland-emergency-law-on-migration-is-a-green-light-for-violence-and-pushbacks-at-the-border/;   https://etias.com/articles/finland-closes-final-russia-border

42. https://www.enisa.europa.eu/sites/default/files/ncss-map/strategies/reports/EE_NCSS_2024_en.pdf

43. https://edition.cnn.com/interactive/2019/05/europe/finland-fake-news-intl/

44. https://www.lrt.lt/en/news-in-english/19/2633078/lithuania-s-new-intelligence-chief-on-russian-sabotage-attacks-and-regional-threats

45. https://www.enisa.europa.eu/sites/default/files/ncss-map/strategies/reports/PL_NCSS_2019_en.pdf

46. https://euroweeklynews.com/2025/11/21/poland-launches-operation-horizon-deploying-10000-troops-today/

47. https://www.defencetoday.com/security/national-security/poland-launches-operation-horizon-to-protect-critical-infrastructure/

48. https://notesfrompoland.com/2025/11/20/poland-launches-military-operation-to-protect-infrastructure-following-russian-rail-sabotage/

49. https://cepa.org/comprehensive-reports/sea-change-nordic-baltic-security-in-a-new-era/;   https://carnegieendowment.org/research/2023/12/from-flooded-meadow-to-maritime-hotspot-keeping-the-baltic-sea-free-open-and-interconnected?lang=en

50. https://www.clustercollaboration.eu/content/baltic-sea-expansion-defense

51. https://www.foi.se/rest-api/report/FOI%20Memo%208504

52. https://news.liga.net/en/politics/news/latvia-wants-to-dismantle-the-railroad-connecting-the-country-with-russia

53. https://www.railbaltica.org/news/rail-baltica-the-line-that-redraws-europes-map/

54. https://cepa.org/article/baltic-defense-getting-new-rail-links-back-on-track/

55. https://www.bbc.com/news/articles/cx2n4r9reejo

56. https://www.euronews.com/my-europe/2025/06/25/rail-baltica-could-be-used-for-defensive-military-purposes

4. Divided Allegiances

1. https://internationalrelations-publishing-files.f1000.com/manuscripts/19255/79c5243e-63d2-421d-b0fd-60aff65243a8_17940_-_agnieszka_bienczyk-missala.pdf?doi=10.12688/stomiedintrelat.17940.1&gtmKey=GTM-MZ2SDMG&immUserUrl=https%3A%2F%2Fwarsaw-proxy.f1krdev.com%2Feditor%2Fmember%2Fshow%2F&otid=9e3c654b-10e7-4556-a1cc-fb442a9f6632&s3BucketUrl=https%3A%2F%2Finternationalrelations-publishing-files.f1000.com&submissionUrl=%2Ffor-authors%2Fpublish-your-research&transcendEnv=cm&transcendId=ef49a3f1-d8c1-47d6-88fc-50e41130631f&numberOfBrowsableCollections=0&numberOf BrowsableInstitutionalCollections=0&numberOfBrowsableGateways=0
2. https://media.un.org/unifeed/en/asset/d334/d3345642
3. https://www.aa.com.tr/en/europe/polish-president-slams-ukraine-nato-allies-for-ignoring-polands-contribution/3626877
4. https://www.voanews.com/a/polish-president-urges-sustained-us-commitment-to-europe-s-security/7860285.html
5. https://ec.europa.eu/eurostat/statistics-explained/index.php?title=Temporary_pro tection_for_persons_fleeing_Ukraine_-_monthly_statistics
6. https://gjia.georgetown.edu/2025/05/21/trump-and-nato/
7. https://www.rferl.org/a/ukraine-baltic-states-weapons-russia/31665474.html
8. https://www.nbcnews.com/world/europe/baltic-states-ask-us-congress-uphold-military-support-rcna230537
9. https://uacua.org/top-15-in-aid-to-ukraine-by-country-gdp/
10. https://kyivindependent.com/lithuania-sends-drones-thermal-imagers-loaders-in-new-ukraine-aid-defense-ministry-says/
11. https://commission.europa.eu/business-economy-euro/economic-recovery/recov ery-and-resilience-facility/country-pages/lithuanias-recovery-and-resilience-plan_en
12. https://www.mod.gov.lv/en/about-us/defense-budget;   https://www.mod.gov.lv/en/drone-coalition-0; https://www.mod.gov.lv/en/drone-coalition-0
13. https://liberalpraktik.fi/baltic-states-contributions-to-ukraine/
14. https://gjia.georgetown.edu/2025/05/21/trump-and-nato/
15. https://sldinfo.com/2023/01/the-perspective-of-president-pavel-shaping-a-way-ahead-for-the-czech-republic/
16. https://kyivindependent.com/czech-ammunition-initiative-for-ukraine-secures-funding-until-september-2025-czech-fm-says/
17. https://kyivindependent.com/czech-initiative-to-deliver-up-to-1-8-million-shells-to-ukraine-pavel-says/
18. https://kyivindependent.com/czech-ammunition-initiative-for-ukraine-secures-funding-until-september-2025-czech-fm-says/
19. https://www.visegradinfo.eu/index.php/national-policy-reports/678-czechia-s-ukraine-aid-under-pressure-as-elections-loom
20. https://www.visegradinfo.eu/index.php/national-policy-reports/678-czechia-s-ukraine-aid-under-pressure-as-elections-loom
21. https://www.praguedaily.news/2025/02/22/poll-majority-of-czechs-reject-further-support-for-ukraine-with-arms-deliveries/

22. https://newsukraine.rbc.ua/analytics/not-only-patriot-how-romania-secretly-helping-1725441851.html; https://www.osw.waw.pl/en/publikacje/analyses/2022-10-14/extremely-cautious-romanias-approach-to-russian-invasion-ukraine

23. https://cepa.org/article/romanian-patriot-missiles-for-ukraine-a-great-day-in-our-lives/

24. https://www.romania-insider.com/romania-military-aid-ukraine-bolojan-august-2025

25. https://www.reuters.com/world/europe/romanian-defense-minister-resigns-pressured-after-ukraine-comment-2022-10-24/

26. https://www.osw.waw.pl/en/publikacje/analyses/2022-10-14/extremely-cautious-romanias-approach-to-russian-invasion-ukraine

27. https://www.bbc.com/news/articles/cn4x2epppeg0

28. https://euromaidanpress.com/2025/03/29/romania-boosts-defense-spending-maintains-ukraines-aid-after-security-council-meeting/

29. https://eualive.net/bulgarias-gas-policy-stalled-diversification-amid-persistent-russian-influence/

30. https://www.oreanda-news.com/en/v_mire/orban-explained-his-refusal-to-support-ukraine/article1576760/

31. https://www.politico.eu/article/ukraine-viktor-orban-volodymyr-zelenskyy-eu-summit-membership-bid/

32. https://europeannewsroom.com/hungary-wants-to-form-a-coalition-against-ukraine-with-other-eu-countries/; https://www.politico.eu/article/so-what-hungary-viktor-orban-shrugs-off-spy-drone-incursion-ukraine/; https://subscriber.politicopro.com/article/eenews/2025/10/07/hungary-clings-to-russian-oil-and-gas-as-eu-and-nato-push-to-cut-supplies-00594959; https://kyivindependent.com/politico-orban-threatens-to-blow-up-eus-ukraine-policy/

33. https://newsukraine.rbc.ua/news/slovakia-announces-new-aid-package-for-ukraine-1759764552.html

34. https://www.mosr.sk/52300-en/v-michalovciach-spustili-prevadzku-na-opravu-ukrajinskej-vojenskej-techniky/

35. https://euromaidanpress.com/2024/10/06/slovak-pro-russain-pm-fico-vows-to-block-ukraines-nato-membership/

36. https://euromaidanpress.com/2024/11/27/slovak-pm-accepts-putins-invitation-to-moscow-victory-day-celebrations-in-2025/

37. https://euromaidanpress.com/2024/10/06/slovak-pro-russain-pm-fico-vows-to-block-ukraines-nato-membership/

38. https://eutoday.net/hungary-and-slovakia-to-block-eu-summit-on-ukraine/

39. https://www.hungarianconservative.com/articles/current/orban-fico-ukraine-nato-energy-security/

40. https://notesfrompoland.com/2024/02/27/differences-over-ukraine-clear-as-polish-hungarian-czech-and-slovak-pms-meet/

41. https://www.eeas.europa.eu/delegations/united-states-america/eu-assistance-ukraine-us-dollars_en?s=253

1. China's Strategic Subordination of Russia

1. https://www.scmp.com/economy/global-economy/article/3250599/chinas-yuan-replaces-us-dollar-euro-russias-primary-foreign-currency-overseas-economic-activity
2. https://www.intereconomics.eu/contents/year/2025/number/2/article/china-russia-cooperation-economic-linkages-and-sanctions-evasion.html
3. https://carnegieendowment.org/russia-eurasia/politika/2023/01/the-risks-of-russias-growing-dependence-on-the-yuan?lang=en
4. https://carnegieendowment.org/russia-eurasia/politika/2024/05/china-russia-yuan?lang=en; https://finance.yahoo.com/news/russia-face-moment-truth-economic-214504181.html
5. https://energyandcleanair.org/july-2025-monthly-analysis-of-russian-fossil-fuel-exports-and-sanctions/
6. https://ideas.repec.org/a/bfr/econot/323.html
7. https://carnegieendowment.org/russia-eurasia/politika/2024/05/china-russia-yuan?lang=en
8. https://defense.info/re-thinking-strategy/2023/05/who-will-compete-with-china-in-central-asia/
9. https://eurasianet.org/one-island-two-countries-a-look-at-how-chinese-russian-relations-are-playing-out-in-the-far-east
10. https://www.gisreportsonline.com/r/russia-far-east/

2. Putin's Beijing Visit: September 2025

1. http://en.kremlin.ru/events/president/news/77914
2. https://thediplomat.com/2025/09/china-mongolia-russia-agreement-on-power-of-siberia-2-could-reroute-energy-trade/
3. https://www.themoscowtimes.com/2025/09/04/what-the-power-of-siberia-2-deal-really-means-for-russia-and-china-a90422; https://carnegieendowment.org/russia-eurasia/politika/2023/04/what-russias-first-gas-pipeline-to-china-reveals-about-a-planned-second-one?lang=en
4. https://www.fmprc.gov.cn/eng/xw/zyjh/202509/t20250904_11702313.html
5. https://blog.ucs.org/robert-rust/what-xi-jinping-tells-his-military-about-taiwan/
6. https://www.cnbc.com/2025/09/03/china-military-day-parade-xi-trump-beijing-us.html
7. https://tass.com/politics/1905505; https://n-ost.org/article/constitution-put-to-test-during-martial-law
8. https://www.the-independent.com/asia/china/trump-putin-sco-summit-modi-ukraine-war-b2818195.html
9. https://focustaiwan.tw/cross-strait/202509030006
10. https://novayagazeta.eu/articles/2025/09/03/german-chancellor-calls-putin-a-war-criminal-for-first-time-en-news
11. https://tass.com/economy/2011827

3. Asian Powers and the Ukraine Conflict

1. https://bulgarianmilitary.com/2025/04/24/japans-secret-satellite-weapon-now-in-ukraines-war-hands/#google_vignette
2. https://www.aerotime.aero/articles/japan-iqps-ukraine-satellite-intel-us-aid-suspension
3. https://united24media.com/latest-news/japan-to-share-sar-satellite-based-data-with-ukraine-for-the-first-time-7763
4. https://theasialive.com/satellite-warfare-japans-iqps-sar-satellites-set-to-strengthen-ukraines-intelligence-arsenal/2025/04/24/
5. https://odessa-journal.com/ukrainian-intelligence-will-receive-satellite-imagery-from-japan
6. https://thediplomat.com/2023/11/south-koreas-quest-to-become-a-global-pivotal-state/
7. https://warontherocks.com/2024/08/south-koreas-growing-role-as-a-major-arms-exporter-future-prospects-in-latin-america/
8. https://asiatimes.com/2024/07/south-koreas-defense-export-growth-a-success-story/
9. https://english.almayadeen.net/news/politics/poland--s--korea-sign-3-year-strate gic-partnership-agreement
10. https://www.koreatimes.co.kr/foreignaffairs/20250306/top-diplomats-of-south-korea-poland-reaffirm-steadfast-commitment-to-advancing-defense-cooperation
11. https://warontherocks.com/2025/04/putting-the-screws-on-the-partnership-between-north-korea-and-russia/
12. https://www.chathamhouse.org/2024/12/north-korea-and-russias-dangerous-part nership
13. https://www.fpri.org/article/2025/03/russia-china-north-korea-relations-obstacles-to-a-trilateral-axis/

4. India and the Ukraine War

1. https://www.pbs.org/newshour/world/putin-and-modi-announce-expansion-of-russia-india-trade-ties
2. https://www.aljazeera.com/news/2025/12/5/putin-modi-kick-off-india-summit-as-trade-us-sanctions-loom-large
3. https://www.eurasiantimes.com/indo-russia-defense-partnership-weathers-all-storms/
4. https://www.institutmontaigne.org/en/expressions/indian-military-dependence-russia
5. https://www.orfonline.org/research/a-defence-link-that-lasts-india-and-russia-in-an-evolving-global-order
6. https://www.researchgate.net/publication/396610502_India's_foreign_policy_re-configuration_from_non-alignment_to_multi-alignment
7. https://warontherocks.com/2020/11/strategic-autonomy-and-u-s-indian-relations/
8. https://www.cfr.org/blog/india-between-superpowers-strategic-autonomy-shadow-pacific-conflict

9. https://valdaiclub.com/a/highlights/state-sovereignty-and-strategic-autonomy-an-indian/
10. https://www.impriindia.com/insights/non-alignment-revisited-strategic/

1. Biden's Ukraine Strategy

1. https://www.cbsnews.com/amp/live-updates/biden-ukraine-military-aid-artillery-ammunition-watch-live-stream-today-2022-04-21/
2. https://www.state.gov/bureau-of-political-military-affairs/releases/2025/01/u-s-security-cooperation-with-ukraine
3. https://www.cfr.org/article/how-much-us-aid-going-ukraine
4. https://www.aljazeera.com/news/2024/12/26/surge-of-weapons-how-much-ukraine-aid-did-biden-approve-after-trump-win
5. https://www.csis.org/analysis/what-ukraine-aid-package-and-what-does-it-mean-future-war
6. https://www.washingtonpost.com/politics/2022/03/10/ukraine-end-game/
7. https://foreignpolicy.com/2022/04/29/russia-ukraine-war-biden-endgame/
8. https://quincyinst.org/2024/10/24/coalition-urges-biden-to-release-congression ally-mandated-unclassified-strategy-for-the-ukraine-war/
9. https://www.rand.org/pubs/commentary/2024/05/bidens-catch-22-in-ukraine.html
10. https://carnegieendowment.org/research/2024/09/russian-military-reconstitution-2030-pathways-and-prospects?lang=en
11. https://carnegieendowment.org/russia-eurasia/politika/2024/12/russia-economy-difficulties?lang=en;   https://thedocs.worldbank.org/en/doc/d5f32e f28464d01f195827b7e020a3e8-0500022021/related/mpo-rus.pdf
12. https://mwi.westpoint.edu/evaluating-us-strategy-for-ukraine-a-pre-postmortem/
13. https://eaworldview.com/2024/06/ukraine-how-2022-peace-talks-failed-halt-russia-invasion/
14. https://www.foreignaffairs.com/ukraine/talks-could-have-ended-war-ukraine
15. https://foreignpolicy.com/2024/05/09/america-ukraine-forever-war-congress-aid/

2. Trump's Ukraine Strategy

1. https://defense.info/defense-decisions/2025/09/the-rhetoric-reality-gap-understanding-trumps-paradoxical-foreign-policy-approach/

1. The State of Play: October 2025

1. https://www.russiamatters.org/news/russia-ukraine-war-report-card/russia-ukraine-war-report-card-sept-17-2025
2. https://www.armed-services.senate.gov/imo/media/doc/general_cavoli_opening_s tatements.pdf
3. https://www.russiamatters.org/news/russia-ukraine-war-report-card/russia-ukraine-war-report-card-feb-26-2025
4. https://www.oryxspioenkop.com/2022/02/attack-on-europe-documenting-equip ment.html; https://www.pravda.com.ua/eng/news/2025/05/18/7512753/

5. https://euideas.eui.eu/the-evolving-rationale-behind-russian-attacks-on-ukraines-energy-infrastructure; https://warontherocks.com/2025/02/the-electricity-front-of-russias-war-against-ukraine/; https://kyivindependent.com/russia-destroys-all-thermal-power-plants-nearly-all-hydroelectric-capacity-in-ukraine-ahead-of-winter-zelensky-says/

6. https://news.un.org/en/story/2025/10/1166016; https://www.reuters.com/business/energy/ukraines-zaporizhzhia-nuclear-plant-now-without-offsite-power-six-days-grossi-2025-09-29/

7. https://www.bbc.com/news/articles/czx020k40560; https://www.themoscowtimes.com/2025/09/26/ukrainian-drones-hit-major-oil-refinery-in-southern-russia-for-second-time-in-a-month-a90634; https://www.kyivpost.com/post/61252

8. https://www.aljazeera.com/news/2025/9/30/russia-ukraine-war-list-of-key-events-day-1314

9. https://www.newsweek.com/nato-scrambles-fighter-jets-russia-12-hour-barrage-ukraine-10793973

10. https://www.themoscowtimes.com/2025/09/26/the-real-losers-of-russias-war-ukraine-europe-and-russia-itself-a90642

11. https://www.nato.int/cps/en/natohq/opinions_237559.htm

12. https://www.nato.int/cps/en/natohq/official_texts_237721.htm

13. https://abcnews.go.com/International/wireStory/russian-forays-nato-airspace-causing-alarm-happening-125954596; https://www.cnn.com/2025/09/25/politics/nato-divided-repeated-russian-incursions-response

14. https://www.lemonde.fr/en/international/article/2025/09/24/alexus-grynkewich-top-nato-commander-in-europe-russia-has-always-tried-to-maneuver-for-advantage-over-the-alliance_6745696_4.html; https://www.koha.net/en/bote/nato-ja-e-ndare-per-reagimin-ndaj-rusise

15. https://news.err.ee/1609807077/defense-minister-nato-was-ready-to-use-force-if-needed-after-russian-airspace-violation; https://kyivindependent.com/nato-was-ready-to-use-of-force-if-needed-tallinn-says/

16. https://kyivindependent.com/calls-to-close-ukraines-skies-return-as-russia-tests-nato-borders/

17. https://reopen.media/en-gb/articles/poland-to-let-forces-target-russian-drones-over-ukraine; https://euromaidanpress.com/2025/09/25/poland-wants-to-shoot-down-targets-over-ukraine-without-nato-and-eu-approval/

18. https://www.euronews.com/2025/09/29/german-def-min-pistorius-russian-federation-is-the-greatest-and-most-immediate-threat-to-n

19. https://www.baltictimes.com/air_defense_should_be_a_nato_priority_this_fall_-_lithuanian_minister/; https://caliber.az/en/post/ukrainian-company-to-launch-solid-rocket-fuel-production-in-denmark; https://www.ukrinform.net/rubric-defense/3949352-britain-to-send-150-artillery-barrels-new-mobile-air-defense-system-to-ukraine.html

20. https://apnews.com/article/europe-ukraine-russia-assets-frozen-loans-plan-91929d8b2263ace7ea452157f19a6222; https://www.nytimes.com/2025/10/01/world/europe/russia-frozen-assets-ukraine-loan.html; https://www.reuters.com/sustainability/boards-policy-regulation/belgium-says-eu-leaders-must-share-risk-use-frozen-russian-assets-ukraine-2025-10-02/

21. https://www.reuters.com/business/energy/slovakia-rejects-eu-plan-phase-out-russ

ian-gas-by-end-2027-2025-05-07/;   https://www.euinsider.eu/news/hungary-and-slovakia-withhold-support-as-eu-proposes-ban-on-russian-gas-and-oil-by-2027

2. Russian Options

1. https://www.themoscowtimes.com/2025/03/29/putin-calls-for-zelenskys-removal-finish-off-ukrainian-troops-a88531
2. https://www.russiamatters.org/news/russia-ukraine-war-report-card/russia-ukraine-war-report-card-oct-1-2025
3. https://www.csis.org/analysis/russias-war-ukraine-next-chapter
4. https://carnegieendowment.org/russia-eurasia/politika/2025/10/russia-refinery-damages?lang=en&utm_source=carnegieemail&utm_medium=email&utm_cam paign=autoemail&mkt_tok=ODEzLVhZVS0oMjIAAAGdSdIOC8AhDPzexcu4u XVw7-lQk3-mBiPIoObwaXULvoV1azGHh6GFamiZcwIcPyfMV23MnCFHpuT8rdUkxrjAjMg NoxiDT4cboECZ3_3tzc5A
5. https://www.newsweek.com/russia-fuel-shortage-ukraine-strikes-10811442
6. https://united24media.com/latest-news/ukraines-drone-blitz-plunges-russia-into-worst-ever-fuel-crisis-knock-out-38-of-refining-capacity-12098
7. https://www.themoscowtimes.com/2025/10/03/ukrainian-drones-hit-major-oil-refinery-chemical-plant-a90706
8. https://www.energyintel.com/00000199-9c4c-d267-a5fb-fccd176b0000
9. https://www.nbcnews.com/world/europe/poland-engages-russian-drones-airspace-first-time-rcna230251
10. https://www.armscontrol.org/act/2025-10/news/nato-downs-russian-drones-over-poland
11. https://www.nato.int/cps/en/natohq/opinions_237559.htm
12. https://www.washingtonpost.com/world/2025/10/01/eu-drone-wall-russia-nato/
13. https://www.cnn.com/2025/09/11/europe/nato-article-four-poland-drones-intl
14. https://www.cnn.com/2025/09/10/europe/putin-nato-poland-what-comes-next-intl
15. https://jamestown.org/program/kremlins-war-economy-driving-recession-in-russias-regions/#:~:text=The%20Kremlin%20spent%208.5%20tril lion,erode%20Russia's%20traditional%20income%20base.
16. https://carnegieendowment.org/russia-eurasia/politika/2025/09/russia-economy-trap?lang=en
17. https://www.piie.com/blogs/realtime-economics/2025/why-russias-economic-model-no-longer-delivers?gad_source=1&gad_campaignid=10567638465&gbraid= 0AAAAADHO67Wob_lOP8q9shNF7Vc98qraR&gclid=CjoKCQjwrojHBhDdARIs AJdEJ_dQuELToQBIICMgXuf9xfZzMik7ovU1pzbqMXWy5t3hovSkMINoiqsaAlau EALw_wcB
18. https://www.csis.org/analysis/how-sanctions-have-reshaped-russias-future
19. https://www.themoscowtimes.com/2025/10/03/russias-private-sector-activity-hits-3-year-low-a90713
20. https://www.businessinsider.com/avtovaz-russia-biggest-carmaker-4-day-work-week-sales-lada-2025-7
21. https://carnegieendowment.org/russia-eurasia/politika/2024/12/russia-economy-difficulties?lang=en

22. https://www.rateinflation.com/inflation-rate/russia-inflation-rate/
23. https://www.wilsoncenter.org/blog-post/risks-russias-two-speed-economy-2025; https://re-russia.net/en/expertise/0238/
24. https://www.dw.com/en/how-russias-mounting-economic-woes-could-force-putins-hand/a-74132905
25. https://www.chathamhouse.org/2025/07/russias-struggle-modernize-its-military-industry
26. https://www.aljazeera.com/news/2025/10/2/putin-warns-of-harsh-response-to-europes-militarisation
27. https://www.kyivpost.com/post/60040
28. https://www.cnbc.com/2025/10/02/russia-will-raise-taxes-to-pay-for-the-ukraine-war.html
29. https://istories.media/en/stories/2025/10/02/putins-war-economy-reaches-limit/
30. https://cepa.org/article/russias-year-of-truth-the-runaway-military-budget/
31. https://carnegieendowment.org/russia-eurasia/politika/2025/09/russia-economy-trap?lang=en
32. https://www.atlanticcouncil.org/blogs/ukrainealert/putins-dream-of-demilitariz ing-ukraine-has-turned-into-his-worst-nightmare/
33. https://www.atlanticcouncil.org/blogs/ukrainealert/putin-begins-2025-confident-of-victory-as-war-of-attrition-takes-toll-on-ukraine/
34. https://www.reuters.com/world/europe/russia-hikes-national-defense-spending-by-23-2025-2024-09-30/
35. https://www.themoscowtimes.com/2025/08/26/from-war-surge-to-slump-is-the-russian-economy-heading-toward-recession-a90354

1. How Nuclear Weapons Shape the War in Ukraine

1. https://defense.info/re-thinking-strategy/2022/07/nuclear-weapons-the-war-in-ukraine-and-russian-geopolitical-maneuver-space/
2. https://defense.info/re-thinking-strategy/2022/04/the-war-in-ukraine-and-the-return-of-the-nuclear-weapons-issue/
3. https://defense.info/re-thinking-strategy/2022/04/the-war-in-ukraine-and-the-return-of-the-nuclear-weapons-issue/
4. https://defense.info/highlight-of-the-week/the-war-in-ukraine-and-the-sanctuar ies-of-rear-support-the-impact-of-nuclear-deterrence/
5. https://defense.info/global-dynamics/2023/01/the-war-in-ukraine-cascading-conse quences/
6. https://defense.info/interview-of-the-week/dr-paul-bracken-on-the-conflict-in-ukraine/

2. The Global War in Ukraine and Globalization

1. https://www.aljazeera.com/news/2025/10/3/how-much-of-europes-oil-and-gas-still-comes-from-russia; https://energyandcleanair.org/publication/eu-imports-of-russ ian-fossil-fuels-in-third-year-of-invasion-surpass-financial-aid-sent-to-ukraine/
2. https://www.cleanenergywire.org/factsheets/liquefied-gas-does-lng-have-place-germanys-energy-future; https://ieefa.org/articles/european-lng-import-terminals-

are-used-less-demand-drops; https://www.spglobal.com/commodity-insights/en/news-research/latest-news/lng/103123-qatar-energys-long-term-lng-contracts-with-european-buyers-likely-to-include-natural-gas-indexation

3. The Global Diffusion of Drone Warfare Skills

1. https://www.aljazeera.com/news/2025/11/8/ukraines-fm-says-over-1400-africans-recruited-to-fight-for-russia-in-war
2. https://www.pbs.org/newshour/show/ukraine-says-russia-is-recruiting-african-mercenaries-to-fight-in-its-war
3. https://adf-magazine.com/2024/07/using-threats-and-false-promises-russia-sends-africans-to-fight-in-ukraine/
4. https://adf-magazine.com/2025/01/russia-tricking-africans-to-fight-war-in-ukraine/
5. https://united24media.com/war-in-ukraine/sold-for-a-passport-russias-recruitment-pipeline-sending-young-africans-into-its-war-8509
6. https://euromaidanpress.com/2024/10/29/international-legion/
7. https://theworld.org/stories/2024/03/13/colombian-army-veterans-join-ukraine-s-army-motivated-financial-need
8. https://english.nv.ua/russian-war/colombian-veteran-fights-for-ukraine-finds-brotherhood-and-purpose-in-the-international-legion-50504716.html
9. https://armyrecognition.com/news/army-news/army-news-2024/colombian-veterans-strengthen-ukrainian-armed-forces-in-its-war-against-russia
10. https://ctc.westpoint.edu/on-the-horizon-the-ukraine-war-and-the-evolving-threat-of-drone-terrorism/
11. https://www.atlanticcouncil.org/blogs/ukrainealert/fpv-drones-in-ukraine-are-changing-modern-warfare/
12. https://www.defensenews.com/unmanned/2025/11/14/how-cartels-are-adopting-drone-tactics-from-ukraine/
13. https://www.atlanticcouncil.org/blogs/ukrainealert/fpv-drones-in-ukraine-are-changing-modern-warfare/
14. https://www.atlanticcouncil.org/blogs/ukrainealert/drone-superpower-ukrainian-wartime-innovation-offers-lessons-for-nato/

www.ingramcontent.com/pod-product-compliance
Lightning Source LLC
Chambersburg PA
CBHW050642270326
41927CB00012B/2838